Symmetry and the Zeros of Riemann's *Zeta* Function

Anthony Lander

Symmetry and the Zeros of Riemann's *Zeta* Function

Two finite mirror image vector series restrict the nontrivial zeros of Riemann's *zeta* function to the critical line and the zeros of its derivative to its right.

Copyright © 2018 Anthony Lander

All rights reserved.

ISBN-13: 978-1986074148

DEDICATION

This monograph is dedicated to anyone who reads and enjoys anything within it. It is especially dedicated to anyone who is able to tell me why the arguments within it are not a proof of the Riemann Hypothesis.

ACKNOWLEDGEMENTS

Anto, Ben, Ellie and Flora have supported and encouraged me, showing interest in this endeavour. Importantly, I acknowledge the patience of my wife Louise who persuaded me to commit my thoughts to paper. Without Louise, this would have been yet another unfinished project, though I suspect she wants it finished so that I then tidy the garage – which is a bigger project altogether.

COVER

The cover shows the pathway of diminishing vectors of Euler's *zeta* for the 901st nontrivial zero. In this work, the vectors highlighted in blue are called ***proximal*** and they run up to a vector called ***kappa***, sitting on a line of reflection $\vec{\psi}$. Vectors after κ form ***distal*** paired pseudo-spirals, here designated $P(\mathcal{R}_r)$, indexed in r and starting with $P(\mathcal{R}_1)$, whose final pseudo-convergence is placed near a vector called ***tau***, which precedes the spiral of divergence. The red vectors, called $\vec{\mathcal{R}}_r$, follow the ***principal-axes*** of the $P(\mathcal{R}_r)$ and have an existence when $r > \kappa$, as seen in the paired pseudo-spirals called $P(\mathcal{M}_m)$ shown underlying the third to fifth vectors of Euler's *zeta*.

The symmetry of Euler's *zeta* is inescapable and the paired spirals have a stability, under changes in the $Re(s)$—which is related to Euler's spiral—and RH follows from the breaking of symmetry. As Pierre-Simon Laplace said…

"Lisez Euler, lisez Euler, c'est notre maître à tous."

CONTENTS

	Acknowledgments	v
	Abstract	1
1	Introduction	1
2	Methods and Resources	5
3	The *Principal-Axis*	28
4	The Breaking of Symmetry	42
5	The Derivatives	64
6	Why the *Zeros* Repel	72
7	The *Beta* Function	91
8	Discussion	95
	Appendix	100
	Notation	123
	References	126

WHY THIS BOOK?

In March 2012 I came across the Riemann Hypothesis and the relationship between the *zeta* zeros and the primes. When I saw how the Fundamental Theorem of Arithmetic was encoded in the *zeta* function equating with Euler's product formula over the primes I was captivated—it is most beautiful. I did not want to enrol on a course or study extensively from books. I wanted to enjoy the thinking and the discovery of beautiful things that I might find on my own. I spent some time contemplating the Harmonic Series, being the infinite sum of the reciprocals of the natural numbers $\zeta(1)$, and trying to work out why it diverged. A week later I had a proof, one I later discovered to be attributed to Nicole Oresme—fantastic fun. I read of Euler solving the Basel problem $\zeta(2)$, being the converging sum of the reciprocals of the squares of the natural numbers equating to $\pi^2/6$. I got nowhere near a proof myself! I was astounded by the richness of $\zeta(2)$. I turned to thinking of Euler's *zeta* with a complex domain.

I saw a pathway of vectors, one for each term of Euler's *zeta* function. I let the complex domain have $\text{Re}(s) < 1$. The early vectors danced around the Argand plane as a function of $\text{Im}(s)$, later vectors were organised into a finite number of paired pseudo-spirals whose axes mimicked the early vectors, and inevitably the pathway entered an infinite spiral of divergence. I found that the final pseudo-convergence preceding divergence, or a related focal point, could be placed at zero for certain values of the $\text{Im}(s)$. I had found a set of numbers remarkably close to the famous nontrivial zeros of Riemann's *zeta*, valid for all *s*. Importantly, pathway symmetry, solely determined by the *zeta* function itself, limited these zeros to the *critical line*. Surely this underlay the Riemann Hypothesis? The paired pseudo-spirals were stable. I extended Dirichlet's alternating and converging *zeta* function to negate less frequent terms and when I did, I found the same symmetries once more. Dirichlet's *beta* behaved similarly. I also found that the mechanism determining the nontrivial zeros precluded Poisson-like randomness. I spent some time looking at cycloid-like behaviour in the pathways and realised that the zeros of the differential of *zeta* all have real part $> 1/2$. Only later did I find that this was equivalent to Speiser's corollary of the Riemann Hypothesis.

I have written this monograph for my own pleasure and I doubt anyone else will read it as I am not a published mathematician. If you do read some small part of this work I hope you enjoy it. If you find mistakes, of which there may be many, please let me know. More importantly, if you can explain why the arguments outlined in this monograph fail to prove Riemann's hypothesis please tell me why.

Anthony Lander   March 2018

Two finite mirror image vector series restrict the nontrivial zeros of Riemann's *zeta* function to the critical line and the zeros of its derivative to its right.

Abstract: Euler's product formula over the primes equates with Euler's *zeta* function to enshrine the Fundamental Theorem of Arithmetic. The famous nontrivial zeros of Riemann's *zeta* function, designated ρ, influence the distribution of the primes. Dirichlet's *eta*, $\eta(s)$ with $s = \sigma + it$, negates alternate terms of *zeta*, converging when $\sigma \leq 1$, and importantly $\eta(\rho) = 0$. Riemann's hypothesis (RH) is that $\text{Re}(\rho) = 1/2$. This work shows that a partial Euler's *zeta* series has a pseudo-convergence at $[t/\pi]$ terms and equates to the difference between paired vector series which run to *kappa* terms with $\kappa = \left[\sqrt{t/(2\pi)}\right]$. The paired vectors have mirror image arguments but their magnitudes differ when $\sigma \neq 1/2$. A modified *eta* series $\eta_l(s)$ is derived with every l^{th} term multiplied by $(1-l)$. If $l > \kappa$, paired vector series to κ terms have a difference that vanishes at the zeros $\eta_l(\rho)$. The *kappa* vectors of the paired series orbit the origin intersecting in a unique way over intervals in t that can host a zero only when $\sigma = 1/2$. Furthermore, it is shown that the vectors of paired series for the derivatives permit zeros only when $\sigma > 1/2$, consistent with Speiser's corollary of RH. The orbiting mechanism precludes Poisson-like randomness and forces the nontrivial zeros to repel, creating an analogy with the pair-correlation function of the eigenvalues of Gaussian Unitary Ensembles. A similar approach proves that Dirichlet's *beta* function satisfies the Generalized Riemann Hypothesis.

---

1. Introduction

1.1. The nontrivial zeros of the Riemann zeta function

Proposition 30 of Book VII of Euclid's Elements, known as Euclid's lemma, states that if p is a prime number and $p|ab$, where a and b are integers then $p|a$ or $p|b$ [1]. This lemma with propositions 31 and 32 give us the Fundamental Theorem of Arithmetic (FTA) that every $n \in \mathbb{N}_{>1}$ is the product of a unique set of primes $n = p_1^{a_1} p_2^{a_2} p_3^{a_3} \ldots p_i^{a_i}$, where $p_1 = 2, p_2 = 3, p_3 = 5, \ldots$. The FTA is enshrined in the *zeta* function equating to Euler's product formula over the primes, $\zeta(x) = \sum_{n=1}^{\infty} n^{-x} = \prod_p (1 - p^{-x})^{-1} = \prod_p (1 + p^{-x} + p^{-2x} + \cdots)$ with $x \in \mathbb{R}$ and is easily seen when the right-hand side is multiplied out; $\zeta(x) = 1 + p_1^{-x} + p_2^{-x} + p_1^{-2x} + p_3^{-x} +$

$p_1^{-x}p_2^{-x} + p_4^{-x} + p_1^{-3x} + p_2^{-2x} + p_1^{-x}p_3^{-x} + \cdots$. Euler's *Zeta* converges when $x > 1$ but Riemann sought $\zeta(s)$ with $s \in \mathbb{C}$ in a form that "remains valid for all s" in analogy to the extension of the factorial function for $s > -1$ in $\Gamma(s+1) = \lim_{n \to \infty} (n+1)^s \frac{1.2.3...n}{(s+1)(s+2)(s+3)...(s+n)}$. Riemann's derivation of $\zeta(s)$ substitutes nx for x in Euler's integral $\Gamma(s+1) = \int_0^\infty e^{-x}x^s dx$ to give $\Gamma(s)n^{-s} = \int_0^\infty e^{-nx}x^{s-1}dx$ for $s > 0$, $n \in \mathbb{N}_1$ which is then summed over n to give $\Gamma(s)\sum_{n=1}^\infty n^{-s} = \int_0^\infty (e^x - 1)^{-1}x^{s-1} dx$ for $s > 1$. Riemann proceeds with a contour integral expressed as $\int_{+\infty}^{+\infty} \frac{(-x)^s}{e^x-1} \frac{dx}{x}$ that runs down a line just above the positive real axis, circles the origin anticlockwise ↶ and then runs back up a line just below the positive real axis. After some manipulation *zeta* is then defined by $\zeta(s) = \frac{\Gamma(s-1)}{2\pi i} \int_{+\infty}^{+\infty} \frac{(-x)^s}{e^x-1} \frac{dx}{x}$ which for real values $s = x$ gives Euler's *zeta* $\zeta(x) = \sum_{n=1}^\infty n^{-x}$ [2].

As the positive integers increase, the primes thin out with $\pi(x)$, the number of primes less than x, being asymptotic to $x/\ln(x)$. This asymptotic distribution is the Prime Number Theorem, conjectured by Gauss in the 1790's and proved independently by Hadamard [3] and de la Vallée Poussin [4] in 1896 using the Riemann *zeta* function. In November 1859 Riemann had published "On the Number of Primes Less Than a Given Magnitude" [5]. Riemann's explicit formula for $\pi_0(x) = \lim_{\varepsilon \to 0} (\pi(x+\varepsilon) + \pi(x-\varepsilon))/2$ is given in terms of $\Pi_0(x)$ which counts the primes and prime powers up to x, counting a prime power p^n as $1/n$ of a prime; $\Pi_0(x) = \sum_{n=1}^\infty n^{-1}\pi_0(x^{1/n})$. The Möbius function $\mu(n)$, is related to the inverse of $\zeta(s)$ being $1/\zeta(s) = \sum_{n=1}^\infty \mu(n)n^{-s}$, which enables the number of primes to be recovered, $\pi_0(x) = \sum_{n=1}^\infty \mu(n)n^{-1}\Pi_0(x^{1/n})$. Riemann's formula then becomes

$$\Pi_0(x) = \text{Li}(x) - \sum_\rho \text{Li}(x^\rho) - \ln(2) + \int_x^\infty (t(t^2-1)\ln(t))^{-1} dt$$

where $\text{Li}(x) = \int_0^x (\ln(t))^{-1} dt$ is the un-offset logarithmic integral and $\{\rho\}$ are the nontrivial zeros.

The Riemann Hypothesis (RH) is that $\text{Re}(\rho) = 1/2 \,\forall\, \rho$ which he did not demonstrate as it was not central to his paper. There is ample evidence [6], but no proof that $\text{Re}(\rho) = 1/2 \,\forall\, \rho$. In the language of the *zeta* function $\sigma = 1/2$ is the *critical line* in the middle of the *critical strip* ($0 < \text{Re}(s) < 1$). In the literature, $\text{Im}(s)$ is either γ or t, we will use t, and let $\sigma = \text{Re}(s)$.

In 1914 Hardy [7] proved the infinitude of $\{\rho\}$ with $\sigma = 1/2$, which we will call $\{\rho_i\}$ indexing those on the *critical line* with i, but Hardy's proof does not preclude there being other real values in the *critical strip* off the line. We will call these unknown hypothetical zeros $\{\rho_u\}$, with u meaning unknown, so that RH is equivalent to the statement $\{\rho_u\} = \emptyset$. Importantly, the Functional Equation, which we will call *Axiom 1*, is

$$\zeta(s) = \Gamma(1-s)(2\pi)^{s-1} 2\sin\left(\frac{\pi s}{2}\right)\zeta(1-s), \tag{1}$$

which encodes symmetries about the *critical line* and about the real axis [2].

In this work, Dirichlet's $\eta(s)$ is modified to define $\eta_l(s)$ in which every l^{th} term of *zeta* is multiplied by $(1-l)$. Our $\eta_l(s)$ shares its nontrivial zeros with Riemann's *zeta* and we use $P(\eta_{l,n}(s))$ to indicate that the n sequential terms of the partial series have been plotted as vectors in $\mathbb{R}^2$; the operator P meaning "pathway of...". A proof of RH follows from symmetry breaking arguments when $P(\eta_l(s))$ is represented by two finite mirror image series.

An important corollary of RH is one published in German in 1935 by Andreas Speiser in *Mathematische Annalen*. Speiser's corollary is that the zeros of the differential of *zeta* lie to the right of the *critical line* and if they lie to the left RH is false [8];

"Auch diese Bedingung ist umkehrbar, so daß der Satz gilt:
Äquivalent mit der Riemannschen Vermutung ist die Behauptung, daß alle nichttrivialen Nullstellen der Ableitung der Zetafunktion rechts von der kritischen Geraden oder auf ihr liegen." Page 520.

This translates as

"Also, this condition is reversible, so that the theorem holds:
Equivalent to Riemann's assumption, is the assertion, that all nontrivial zeros of the derivative of the Zeta function lie on, or to the right of the critical line. "

A geometric proof of this corollary is presented here and contributes to a second proof of RH.

1.2. The Generalized Riemann Hypothesis and the Dirichlet beta function

Reflecting on the Leibniz-Madhava series $\pi/4 = 1 - 1/3 + 1/5 - 1/7 \ldots$, which is a Dirichlet L-function, one is instantly inquisitive about the zeros of the alternating and converging series

$$\beta(s) = 1^{-s} - 3^{-s} + 5^{-s} - 7^{-s} + 9^{-s} \ldots \text{ using } s = \sigma + it$$

with $\sigma, t \in \mathbb{R}$ and $i = \sqrt{-1}$. \tag{2}

The Generalized Riemann Hypothesis (GRH) makes a statement similar to RH about the real part of the domain of the zeros of Dirichlet L-functions being 1/2. For $\beta(s)$ a set of zeros for low values of t are easy to calculate to within a few decimal places and all seem to have $\sigma = 1/2$. Not surprisingly, the set $\{\text{Im}(s): \beta(s) = 0\} \neq \{\text{Im}(\rho_i)\}$. In this work, the vector pathway to convergence of $\beta(s)$ in $\mathbb{R}^2$ is designated $P(\beta(s))$ and analysis of its geometry yields a symmetry breaking argument which proves that the *beta* function obeys GRH.

1.3. Quantum Chaos and the eigenvalues of Ensembles of Random Matrices

Deterministic classical mechanical systems can be exponentially sensitive to initial conditions creating pseudo-random distributions with near periodic outputs which are described as chaotic. In contrast, Quantum Mechanics is probabilistic and not deterministic. The anxiety that this creates is that it is not immediately obvious as to how a quantum mechanical output can be exponentially sensitive to the initial conditions in the classical limit, which the correspondence principle requires [9].

Neutron capture by atomic nuclei is the mechanism for the cosmic synthesis of heavy elements which cannot be formed by nuclear fusion. After a brief pause, atomic nuclei which absorb neutrons can release quantized radioactive energy in spectra which physicists model in statistical terms with Ensembles of Random Hermitian matrices. The Gaussian Unitary Ensemble (GUE) is such a collection and is used to model random Hamiltonians which have the attraction of lacking the symmetry of time reversal. In passing, we note two other Ensembles which are reversible; the Gaussian Orthogonal Ensemble (GOE) and the Gaussian Symplectic Ensemble (GSE). However, it is the GUE which is relevant to the Riemann *zeta* function since the eigenvalues of the Hamiltonian are considered to be analogous to the nontrivial zeros of the *zeta* function.

The normalised intervals between pairs of nontrivial zeros behave like the intervals between pairs of eigenvalues of random Hermitian matrices in the Gaussian Unitary Ensemble. Of relevance to this work is the observation that neither the zeros of the *zeta* function nor the eigenvalues have Poisson-like randomness, with immediate neighbours repelling one another, but the mechanism of repulsion is not understood [10, 11, 12]. The proof of RH presented here has two functions which orbit the origin and create nontrivial zeros when they meet. Repulsion between the nontrivial zeros is a necessary consequence of this orbiting mechanism and follows directly from this proof of RH.

The final vectors of the paired finite series are designated *kappa*. When Re(s) = 1/2 the pathways of the paired vectors are mirror images about a line of reflection designated *psi*, $\vec{\psi}$. The line of reflection and the *kappa* vectors orbit the origin as Im(s) rises. These three rotate at different rates and only when the *kappa* vectors intersect at *psi*, can an imaginary interval host a zero. Immediately following a nontrivial zero, the faster *kappa* vector rotates before catching up with its slower partner to meet it at *psi* once more to generate the next zero. For this reason, the nontrivial zeros appear to repel, and the tails of the distribution, in comparison with a simple probabilistic model, flatten. The resemblance with the pair-correlation function of the Gaussian Unitary Ensemble immediately follows.

2. Methods and Resources

2.1. Methodological approach

This work started with two premises; a naïve premise and a sober premise. The naïve premise was that an understanding of the vector pathway of $\eta_l(s) = (1 - l^{1-s})\zeta_\infty(s)$, when combined with the Functional Equation, *Axiom 1*, would prove RH, or establish the potential for zeros to exist off the *critical line*. Cognisant of history it was anticipated that the naïve premise would fail. The sober premise was that an understanding of that failure would be informative.

2.2. Resources

Microsoft Excel 2010 with Visual Basic and GraphPad Prism 7 were used to illustrate the naïve premise. Calculations were assisted with data from a published set of nontrivial zeros [13].

2.3. Terminology and notation

Euler's *zeta* is typically formulated as $\zeta(x) = \sum_{n=1}^{\infty} n^{-x}$ with $x \in \mathbb{R}$, whilst $\zeta(s)$ with $s \in \mathbb{C}$ is the analytically continued Riemann *zeta* function valid for all s. To distinguish these two we give Euler's *zeta* a subscript, in $\zeta_\infty(x)$, so that it can take a complex domain in $\zeta_\infty(s) = \sum_{m=1}^{\infty} m^{-s}$, which diverges when Re(s) ≤ 1, but can be tamed in $\zeta_n(s) = \sum_{m=1}^{n} m^{-s}$ by limiting n. We retain $\zeta(s)$, without a subscript, for Riemann's *zeta* valid for all s.

The italicised "zero" is used to mean a nontrivial zero ρ. We first address *zeta*, *eta* and RH leaving *beta* and the GRH for Section 7 page 91. Four diverging series $\zeta_n(s)$, $h_{l,r}(s)$, $\bar{h}_{l,r}(s)$ and $\ell_{l,n}(s)$ and the converging series $\eta_{l,n}(s)$ are described. If $f_{l,n}(s)$ is a vector series to n terms influenced by $l \in \mathbb{N}_1$, then $P(f_{l,n}(s))$ is used to mean the pathway of sequential vectors representing the terms of that series when plotted in $\mathbb{R}^2$. In this way $P(f_{l,n}(s))$ is richer than $f_{l,n}(s)$ which on its own identifies a point in the plane or implies a vector to that point. The richness comes from characterizing how pathway geometry, under *Axiom 3* (page 23), limits the possibilities for $P(f_{l,n}(s))$ as σ changes.

If $f_{l,n}(s)$ is used as a vector this is implied by context or made explicit in the notation $\vec{f}_{l,n}(s)$. Each series has a specific index using different fonts/scripts; care is needed to avoid confusion. The four divergent vector pathways are $P(\zeta_n(s))$, $P(h_{l,r}(s))$, $P(\overline{h}_{l,r}(s))$ and $P(\ell_{l,n}(s))$, which summate using the indices m, r, r and n respectively to final terms of n, r, r and n. The vectors of these series are $\vec{m}$, $\vec{\mathcal{R}}_r$, $\vec{\overline{\mathcal{R}}}_r$ and $\vec{\mathcal{L}}_n$ respectively. By way of example, the converging $P(\eta_{l,n}(s))$ has index m, runs to n terms and summates vectors designated $\vec{m}$.

Individual vectors can be added to a pathway such that they run sequentially. Where appropriate we consider $P(f_{l,n}(s))$ to imply $\vec{f}_{l,n}(s)$ such that $P(f_{l,n}(s)) = P(g_{l,n}(s))$ means $\vec{f}_{l,n}(s) = \vec{g}_{l,n}(s)$ but this does not mean the pathways are identical, it does not mean $P(f_{l,n}(s)) \equiv P(g_{l,n}(s))$.

The diverging pathways have *proximal* and *distal* parts separated by a vector at a term *kappa*, κ. The *distal* part of each divergent pathway passes through a finite number of paired pseudo-spirals before entering an infinite spiral of divergence. In a diverging series the term corresponding to the vector preceding the spiral of divergence is designated *tau*, τ.

The converging $P(\eta_{l,n}(s))$ has *proximal* and *distal* parts separated by a vector at the term κl. The *distal* part of the converging $P(\eta_{l,n}(s))$ passes through a finite number of paired pseudo-spirals, designated $P(\mathcal{R}_r)$, and needs no *tau* vector as true convergence lies within the final pseudo-spiral. In this way $P(\eta_{l,\infty}(s)) = P(\eta_{l,\kappa l}(s)) + \sum_{r=\kappa}^{r=1} P(\mathcal{R}_r)$.

For $P(\eta_{l,n}(s))$ and $P(\zeta_n(s))$ the terms *proximal* and *distal* are also used within the paired pseudo-spirals, already designated $P(\mathcal{R}_r)$; the $P(\mathcal{R}_r)$ being the pathway of a set $\mathcal{R}_r$ of sequential $\vec{m}$. *Proximal* describes the clockwise unwinding pseudo-spiral and *distal* the anticlockwise pseudo-spiral which winds inwards, the two spirals are separated at an *inflection point* (located with an integral $q_{\bar{r}}$). The term "pseudo-convergence" refers to the transition between neighbouring $P(\mathcal{R}_r)$ (located with an integral q_r), and each pseudo-convergence implies a "focal point" within the pseudo-spiral. The focal points in the spirals of the $P(\mathcal{R}_r)$ form the ends of a vector $\vec{\mathcal{R}}_r$ which represents the *principal-axis* of the $P(\mathcal{R}_r)$. Importantly, the $\vec{\mathcal{R}}_r$ have an existence independent of the $P(\mathcal{R}_r)$ allowing r to rise without limit.

When a function has more than one parameter $f(a, b, c \ldots)$ and only one is to be varied then that function is denoted with a solitary variable in its brackets e.g. $f(b)$ implies a and c are fixed. The function $\eta_l(s)$, can be considered as $\eta_l(\sigma)$ with t fixed, or $\eta_l(t)$ with σ fixed; this is useful since curves with one parameter fixed are informative. Curves in $\eta_l(\sigma)$ being especially informative in the context of the limitations of the behaviour of $P(\eta_l(\sigma))$, limitations consequent upon *Axiom 3* (page 23).

The known *zeros* are the set $\{\rho_i\}$, each having $\text{Re}(\rho_i) = 1/2$ and $t_i = \text{Im}(\rho_i)$. An unknown nontrivial zero off the *critical line* is designated ρ_u with $\text{Re}(\rho_u) \neq 1/2$ and $t = t_u$ with $t_u \notin \{t_i\}$. The Functional Equation, *Axiom 1* in Equation 1, demands that if $\zeta(\sigma_\alpha + it_u) = 0$ then $\zeta(\sigma_\beta - it_u) = 0$ when the restriction $\sigma_\beta = 1 - \sigma_\alpha$ applies. We limit σ_α to the left of the *critical line* and σ_β to the right. Since the sign of t can be ignored without weakening our conclusions we consider that if there is a $\rho_{u_\alpha} = \sigma_\alpha + it_u$ then there is a $\rho_{u_\beta} = \sigma_\beta + it_u$, see Figure 1.

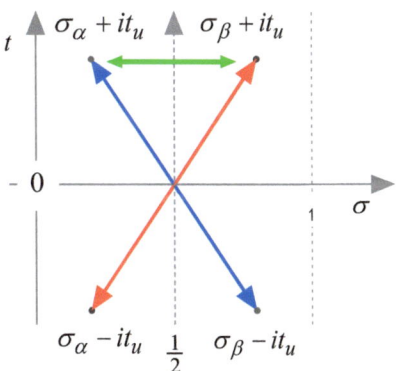

Figure 1. The *critical strip* and *critical line* with four related hypothetical *zeros*. The Functional Equation follows the red and blue arrows. We limit $\sigma_\alpha < 1/2 < \sigma_\beta$. The green line maps $\rho_{u_\alpha} = \sigma_\alpha + it_u$ to $\rho_{u_\beta} = \sigma_\beta + it_u$ which is equivalent to the Functional Equation being agnostic to the sign of t.

From here we only need $t > 0$. We let RH mean the Riemann Hypothesis $\text{Re}(\rho) = 1/2$ for all ρ. The set $\{\rho\} = \{\rho_i\} \cup \{\rho_u\}$ and so RH $\Leftrightarrow \{\rho\} \equiv \{\rho_i\}$ and RH $\Leftrightarrow \{\rho_u\} = \emptyset$. We let $\mathbb{O} = \{1, 3, 5, 7 \ldots\}$, $\mathbb{E}_2 = \{2, 4, 6 \ldots\}$, $\mathbb{N}_0 = \{0, 1, 2, 3 \ldots\}$, $\mathbb{N}_1 = \{1, 2, 3, \ldots\}$, $\mathbb{N}_{>1} = \{2, 3, 4 \ldots\}$ and $\mathbb{Z} = \{\ldots -3, -2, -1, 0, 1, 2, 3 \ldots\}$, $\mathbb{Q}$ the rational numbers, $\mathbb{R}$ the reals and $\mathbb{C}$ the complex numbers.

We map $m^{-s} \in \mathbb{C}$ into $\mathbb{R}^2$ as follows; $m^{-s} = m^{-\sigma}m^{-it}$ and $m^{-it} = e^{\ln(m^{-it})} = e^{-it\ln(m)}$ and so by Euler's Identity $m^{-s} = m^{-\sigma}(\cos(t\ln(m)) - i\sin(t\ln(m)))$ which we plot in $\mathbb{R}^2$. The notation $[x]$ means the nearest integer to $x \in \mathbb{R}$, with floor $\lfloor x \rfloor$ and ceiling $\lceil x \rceil$ functions meaning the nearest integers below and above x respectively. The notation $[\![x]\!]$ means the nearest element to x in an ordered set; if the set were $\mathbb{E}_2$ then $4 = [\![3.1]\!]$. Additional objects appear in the text and in **Notation** page 123.

2.4. Preliminaries

This section introduces five series $\zeta_n(s)$, $\eta_{l,n}(s)$, $\ell_{l,n}(s)$, $h_{l,r}(s)$ and $\bar{h}_{l,r}(s)$ and the associated numbers $\kappa, \tau, q, l \in \mathbb{N}_1$ and $\check{\kappa}, \psi \in \mathbb{R}$, a curve $Y_r(u)$ and the sets R_l, χ_l and $\mathcal{R}_r$.

1. A partial Euler's *zeta* series and its vector representation in $\mathbb{R}^2$ are defined as;

$$\zeta_n(s) = \sum_{m=1}^{n} m^{-s},$$

and $P(\zeta_n(s)) \equiv \sum_{m=1}^{n} \vec{m}$ plotted sequentially in $\mathbb{R}^2$

with $|\vec{m}| = m^{-\sigma}$ and $\arg(\vec{m}) = -t\ln(m)$. (3)

2. The orientation of neighbouring $\vec{m}$ and $\overrightarrow{m+1}$ in $P(\zeta_n(s))$ has, looking forwards, a vector $\vec{m}$ of length $m^{-\sigma}$ neighbouring a vector at $(m+1)$ of length $(m+1)^{-\sigma}$ lying at an angle of $\Delta\vartheta = t\ln(1 + 1/m)$ with a reduced form $\Delta\theta = \Delta\vartheta - 2\pi r$ where $r \in \mathbb{N}_1$, with $r = [(t/2\pi)\ln(1 + 1/m)]$ so that $-\pi < \Delta\theta \leq \pi$. An $r < \kappa$ (kappa is defined in *Preliminary 4*) is then the index of paired pseudo-spiral structures designated $P(\mathcal{R}_r)$. Figure 2(a) (opposite on page 9), illustrates a pathway a little after the *inflection point* in a $P(\mathcal{R}_r)$ with $r = 1$, showing $\Delta\theta = \Delta\vartheta - 2\pi$ at a value $\Delta\theta$ a little less than 0. As m rises, the pathway traces irregular polygons with a reducing number of sides before triangles are formed (Figure 2(b)) which get smaller and narrower as vectors repeatedly double-back on themselves.

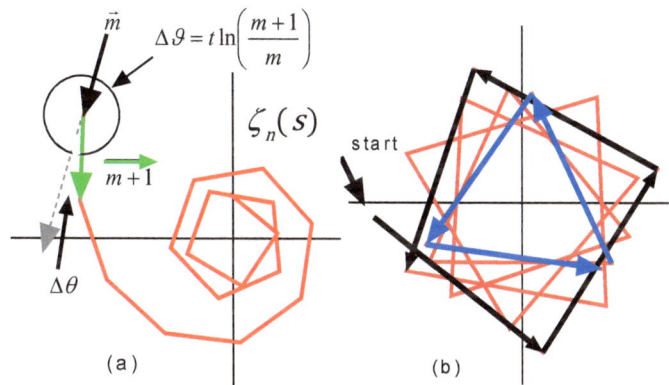

Figure 2. (a) The final converging pseudo-spiral in P($\zeta_n(s)$) beginning when $\Delta\vartheta$ is $< 2\pi$ and $\Delta\theta$ is < 0. (b) Later vectors with a quadrilateral (in black), first appearing when $\Delta\theta < -\pi/2$, and a triangular path, first appearing when $\Delta\theta < -2\pi/3$ (in blue).

Figure 2 illustrates the final pseudo-spiral of P($\zeta_n(s)$) after the *inflection point* in P($\mathcal{R}_1(s)$) when $\Delta\vartheta$ is a little less than 2π and $\Delta\theta$ is a little less than 0. When $\Delta\vartheta \approx \pi$ and $\Delta\theta \approx -\pi$ the *tau* vector (see below) is found, and thereafter the final spiral of divergence if $\sigma \leq 1$.

3. If in P($\zeta_n(s)$) the r associated with $t\ln(m/(m-1))$ and that associated with $t\ln((m+1)/m)$ differ by more than 1 then that m is in the *proximal* pathway. If consecutive $\vec{m}$ share the same r, then they are in the same P($\mathcal{R}_r$). Similar relations apply for $h_{l,r}(s)$ and $\ell_{l,n}(s)$.

4. *Kappa and kappa dot.* In P($\zeta_n(s)$) the *proximal* pathway is a set of individual vectors lacking simple structure because the $\Delta\vartheta$ associated with $t\ln(m/(m-1))$ and that associated with $t\ln((m+1)/m)$ differ by more than 2π. In contrast, the *distal* pathway has predictable structures because the $\Delta\vartheta$ associated with $t\ln(m/(m-1))$ and that associated with $t\ln((m+1)/m)$ differ by less than 2π. If $\mu \in \mathbb{R}$ replaces m we seek μ in $t\ln(\mu/(\mu-1)) - t\ln((\mu+1)/\mu) = 2\pi$ or $t\ln(\mu^2/(\mu^2-1)) = 2\pi$ from which $\mu = (1 - e^{-2\pi/t})^{-1/2}$. A modest refinement, which is important at very low values of t, gives *kappa*

$$\kappa = \left[\frac{1}{2}\left((e^{2\pi/t} - 1)^{-1/2} + (1 - e^{-2\pi/t})^{-1/2}\right)\right]. \quad (4)$$

Kappa's two terms embrace $\sqrt{t/(2\pi)}$ since

$$(1 - e^{-2\pi/t})^{-1/2} > \sqrt{t/(2\pi)} > (e^{2\pi/t} - 1)^{-1/2}$$

with $\frac{1}{2}\left(\left(e^{2\pi/t} - 1\right)^{-1/2} + \left(1 - e^{-2\pi/t}\right)^{-1/2}\right) \approx \sqrt{t/(2\pi)}$ at high values of t when $\kappa = \left[\sqrt{t/(2\pi)}\right]$.

5. A residual, *kappa dot* written $\dot{\kappa} \in \mathbb{R}$, with $-1/2 \leq \dot{\kappa} \leq 1/2$ is simply,

$$\dot{\kappa} = \frac{1}{2}\left(\left(e^{2\pi/t} - 1\right)^{-1/2} + \left(1 - e^{-2\pi/t}\right)^{-1/2}\right) - \kappa. \tag{5}$$

6. *Kappa dot* has a relationship with $x \in \mathbb{R}$, $0 < x \leq 1$ being the fractional intersection of the *kappa* vectors $\vec{\mathcal{L}}_\kappa$ and $\vec{\mathcal{R}}_\kappa$ at the *zeros* (see below) and is captured in $x = f(\dot{\kappa})$.

7. Four divergent series $\zeta_n(s)$, $\ell_{l,n}(s)$, $h_{l,r}(s)$ and $\bar{h}_{l,r}(s)$ each have *proximal* and *distal* parts separated at a term designated *kappa*. *Proximal* and *distal* parts of $P(\zeta_\tau(s))$ are separated by a $\vec{m}$ for which $m = \kappa$, and which is designated $\vec{\kappa}$. There are similar *kappa* vectors $\vec{\mathcal{L}}_\kappa$ and $\vec{\mathcal{R}}_\kappa$ at the end of the series $\ell_{l,\kappa}(s)$ and $h_{l,\kappa}(s)$ respectively. These are described below.

8. An $l \in \mathbb{N}_1$ identifies every l^{th} term in a series allowing a different rule to apply to the magnitude or argument of the lth vectors. Where appropriate, Euler's *zeta* is assigned a value of $l = 1$, since every term is treated the same as its neighbours.

9. We derive $\eta_l(s)$, valid in the *critical strip*, which shares a relationship with Riemann's *zeta* $\zeta(s)$ "valid for all s". Starting with $\zeta_\infty(s) = \sum_m^\infty m^{-s}$, the modified Dirichlet *eta* function is derived through multiplication by l^{1-s} which is then subtracted from $\zeta_\infty(s)$ to give a series with every l^{th} term multiplied by $(1 - l)$;

$$\eta_l(s) = (1 - l^{1-s})\zeta_\infty(s). \tag{6}$$

For example if $l = 5$, $\eta_5(s) = 1 + \frac{1}{2^s} + \frac{1}{3^s} + \frac{1}{4^s} - \frac{4}{5^s} + \frac{1}{6^s} + \frac{1}{7^s} + \frac{1}{8^s} + \frac{1}{9^s} - \frac{4}{10^s}\ldots$. It can be seen that $\eta_l(s)$ shares its *zeros* with Riemann's *zeta* $\zeta(s)$. Though *zeta* has a pole at $s = 1$ our $\eta_l(s)$ has a domain when $\sigma = 1$, which is helpful.

10. A partial series $\eta_{l,n}(s)$ is defined as

$$\eta_{l,n}(s) = \sum_{m=1}^{n} jm^{-s}$$

$$= \sum_{m=1}^{n} jm^{-\sigma}(\cos(t\ln(m)) - i\sin(t\ln(m))),$$

with $j = 1$ when $l \nmid m$
and $j = (1 - l)$ when $l \mid m$. \hfill (7)

The pathway $P(\eta_{l,n}(s))$ is a sequential plot of the n vectors, indicated $\vec{m}$, as follows;

$$P(\eta_{l,n}(s)) \equiv \sum_{m=1}^{n} \vec{m} \text{ plotted in } \mathbb{R}^2 \text{ with}$$

$|\vec{m}| = jm^{-\sigma}$ and $\arg(\vec{m}) = -t\ln(m) + \phi$,
with $j = 1$ and $\phi = 0$ when $l \nmid m$
and $j = (l - 1)$ and $\phi = \pi$ when $l \mid m$. \hfill (8)

11. In $\eta_{l,n}(s)$ an m for which $m = \kappa l$, separates *proximal* values of m, for which $m < \kappa l$, from *distal* values of m for which $m > \kappa l$.

12. The *distal* pathway has a number of paired pseudo-spirals $P(\mathcal{R}_r)$ since neighbouring $\vec{m}$ share the same r. Each $P(\mathcal{R}_r)$, of $P(\zeta_n(s))$ or of $P(\eta_{l,n}(s))$, has a *principal-axis* running between focal points and equating to a vector $\vec{\mathcal{R}}_r$, whose argument is stable under changes in σ. When seen as m rises, a $P(\mathcal{R}_r)$ has a clockwise ↻ unwinding pseudo-spiral, a straighter region that crosses the *principal-axis* at $3\pi/4$, and an anticlockwise ↺ pseudo-converging pseudo-spiral.

13. The series $\zeta_n(s)$ has a transition when $m = \tau$ at a vector $\vec{m} = \vec{\tau}$ in $P(\zeta_n(s))$ which signals the end of the final pseudo-convergence and the start of spiraling divergence. *Tau* is a function of t alone in Euler's zeta and in the series $\ell_{l,n}(s)$, which is described below. *Tau* is also a function of t and l for the diverging series $h_{l,r}(s)$ and $\bar{h}_{l,r}(s)$, both of which summate series of vectors also designated $\vec{\mathcal{R}}_r$ and described below. There are two related formulations for *tau* which follow directly from the geometry of the pathways;

$$\tau = \left[\frac{1}{2}\left(\left(e^{\frac{\pi}{t}} - 1\right)^{-1} + \left(1 - e^{-\frac{\pi}{t}}\right)^{-1}\right)\right] \approx \left[\frac{t}{\pi}\right]$$

for Euler's *zeta* $\zeta_\tau(s)$ and $\ell_{l,\tau}(s)$, \hfill (9) and $\tau = \left[\left(e^{\frac{\pi}{tl}} - 1\right)^{-1} + \left(1 - e^{-\frac{\pi}{tl}}\right)^{-1}\right] \approx \left[\frac{2tl}{\pi}\right]$

if $l \geq 2$ for $h_{l,\tau}(s)$ and $\bar{h}_{l,\tau}(s)$. (10)

The approximations $[t/\pi]$ and $[2tl/\pi]$ suffice for discussing RH and indeed they are poor only at very low values of t.

14. There will be a trivial zero for $\eta_l(s)$ when $\sigma = 1$ if $t = 2\pi k/\ln(l)$ with $k \in \mathbb{N}_1$, and these will be unique, excluding those generated by raising l to l^a with $a \in \mathbb{N}_{>1}$. These trivial zeros play a role in later arguments and a proof of their existence follows in *Lemma (1)*.

Lemma (1). There are trivial zeros $\eta_l(1 + it) = 0$ if $t = 2\pi k/\ln(l)$ and $k \in \mathbb{N}_1$.

Proof: Letting $a, b \in \mathbb{N}_1$ we see that $b^{-1} - \sum_{a=1}^{\infty} \frac{l-1}{bl^a} = 0$ in two ways. (A) It is easy to see that $1 - \frac{1}{2} - \frac{1}{4} - \frac{1}{8} \ldots = 0$ and likewise $1 - \frac{2}{3} - \frac{2}{9} - \frac{2}{27} \ldots = 0$ and so $1 - \frac{l-1}{l} - \frac{l-1}{l^2} - \frac{l-1}{l^3} \ldots = 0$ yielding $b^{-1} - \frac{l-1}{bl} - \frac{l-1}{bl^2} - \frac{l-1}{bl^3} \ldots = 0$. (B) Let $s = \frac{1}{l} + \frac{1}{l^2} + \frac{1}{l^3} + \frac{1}{l^4} + \frac{1}{l^5} + \cdots$ so that $ls = 1 + \frac{1}{l} + \frac{1}{l^2} + \frac{1}{l^3} + \frac{1}{l^4} + \cdots$ giving $ls - 1 = s$ and $\frac{1}{l-1} = s$ which is $\frac{1}{l-1} = \sum_{a=1}^{\infty} \frac{1}{l^a}$ from which it follows that $1 - \sum_{a=1}^{\infty} \frac{l-1}{l^a} = 0$ and hence $b^{-1} - \sum_{a=1}^{\infty} \frac{l-1}{bl^a} = 0$. We also see that $b^{-\sigma} - \sum_{a=1}^{\infty} \frac{l-1}{(bl^a)^\sigma} \neq 0$ if $\sigma \neq 1$. We restrict $a, b \in \mathbb{N}_1$ such that $l \nmid b$ and $a \neq b$ and consider terms in $\eta_{l,n}(1 + it)$ which are positive when $m = b$ and negative when $m = bl^a$, since $l|m \Rightarrow j = 1 - l$ (Equation 7, page 11). The $\vec{m}$, for which $m = bl^a$ have arguments $t\ln(bl^a) = ta\ln(l) + t\ln(b)$ into which we substitute $\ln(l) = 2\pi k/t$ to give $t\ln(bl^a) = 2\pi ka + t\ln(b)$. With $2\pi ka \equiv 2\pi$ we now see that

$$\eta_l(\sigma + it) = \lim_{b \to \infty} \sum_{b=1}^{\infty} \left(b^{-\sigma} e^{it\ln(b)} - \sum_{a=1}^{\infty} \left(\frac{l-1}{(bl^a)^\sigma}\right) e^{i2\pi} e^{it\ln(b)}\right)$$

$$= e^{it\ln(b)} \lim_{b \to \infty} \sum_{b=1}^{\infty} \left(b^{-\sigma} - \sum_{a=1}^{\infty} \left(\frac{l-1}{(bl^a)^\sigma}\right)\right),$$

and so $\eta_l(1 + it) = 0 + 0i$ when $\sigma = 1$ and $t = 2\pi k/\ln(l)$.

15. Three examples for $P(\eta_2(\rho_i))$ are shown in Figure 3, omitting the first vector $\vec{1}$ for clarity. The *proximal* pathways before an $\vec{m}$ for which $m = \kappa l$ have a smaller number of much larger vectors than the *distal* pathways which have very many small vectors forming the $P(\mathcal{R}_r)$.

The P($\mathcal{R}_1$) all have a *principal-axis* of magnitude $\sqrt{l}$, shown sitting on a circle of radius $\sqrt{2}$. The parameters for Figure 3 are, $\sigma = 1/2$ $t_i = 59.347$ (red) $\kappa l = 6$, $t_i = 376.324$ (blue) $\kappa l = 16$, and $t_i = 74920.8275$ the $100{,}000^{th}$ zero in black with $\kappa l = 218$.

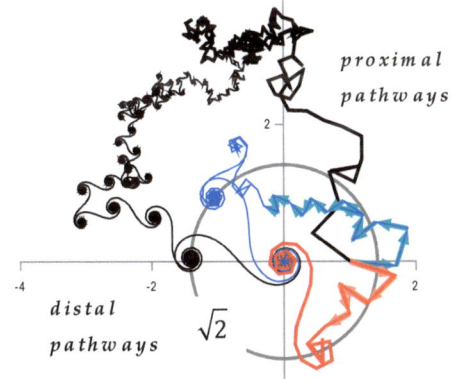

Figure 3. P($\eta_2(\rho_i)$) with $\sigma = 1/2$ for three t_i. The t_i were chosen so that the highlighted *proximal* vectors, preceding the κl vectors, lie predominantly on the right-hand side.

16. When $l > 2$ the P($\mathcal{R}_r$) of P($\eta_l(s)$) contain near-regular polygons and star-polygons and the geometry between neighbouring paired pseudo-spirals for a P($\mathcal{R}_r$) differ from those in P($\eta_2(s)$). Convergences for $l = 5$ and $l = 23$ are shown in Figure 4. As l rises the final pseudo-spiral occurs at higher values of m and the final pseudo-spiral contains near-regular polygons with one side missing.

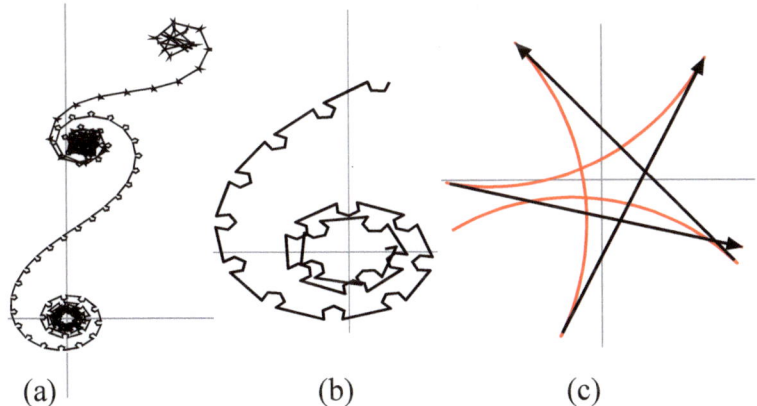

(a)   (b)   (c)

Figure 4. (a) The final paired pseudo-spirals of a P($\eta_5(s)$).
(b) Detail from a final pseudo-spiral showing near-regular pentagons, each missing one side and lying between neighbouring 5th vectors.
(c) With $l = 23$, 22 vectors precede convergence and resemble smooth arcs (red), followed by the larger returning vectors (black); the curvature of each arc gradually reduces as m rises.

17. In $\eta_l(s)$ when $l > 2$ there are near-regular l sided polygons and near-regular l sided star-polygons along $P(\eta_l(s))$. The polygons lie between the $\vec{m}$ where $l|m$. The rules applying when $l > 2$ generalise to $l = 2$. The index r falls as m rises. A function $m(q)$ locates the pseudo-convergences between neighbouring $P(\mathcal{R}_r)$ and the *inflection points* within a $P(\mathcal{R}_r)$ by locating important geometric changes which take place in the polygons. Appendix A1 examines these geometries. The difference in arguments between neighbouring $\vec{m}$ is $t\ln((m+1)/m)$ looking forward and $t\ln(m/(m-1))$ looking backwards. Important pathway changes occur when neighbouring differences lie immediately either side of specific integral multiples of π/l. The integer $q \in \mathbb{N}_1$ identifies significant $\vec{m}$ with $m(q) \approx [tl/(q\pi)]$,

$$m(q) = \left[\frac{1}{2}\left(\left(e^{\frac{q\pi}{tl}} - 1\right)^{-1} + \left(1 - e^{-\frac{q\pi}{tl}}\right)^{-1}\right)\right]. \tag{11}$$

18. Let Q_q contain all points in $P(\eta_l(s))$ between $m = m(q+1)$ and $m = m(q)$. We should not include $q = 0$, since Equation 11 is undefined, but we can give meaning to Q_0 by briefly allowing $q \in \mathbb{R}$ when $0 < q \le 1$ recognising the $\lim_{q \to 0}\left(e^{q\pi/tl} - 1\right)^{-1} = \infty$. Conscious of these limitations temporary meaning is given to Q_q as a set;

$$Q_q = \{\vec{m} : m(q) > m \ge m(q+1)\} \quad \text{with } q \in \mathbb{N}_0. \tag{12}$$

19. The set $M_l = \{m : m = m(q)\}$ contains important points in the pathway $P(\eta_l(s))$. The structural unit Q_q may be a substantial part of a paired pseudo-spiral $P(\mathcal{R}_r)$ or a region that makes no progress in $\mathbb{R}^2$ but rather contains oscillations about a pseudo-convergence.

20. We now consider $\eta_{l,n}(s)$ ending near the focal point of a pseudo-spiral with $m \in M_l$. The end point of the final $\vec{m}$, with $n = m(q)$ in $\eta_{l,n}(s)$, will not best represent the focal point. This is evident in pseudo-spiral pairs at low values of t and m, and is especially evident when $l \gg 2$, since $|\vec{m}|$ when $l|m$ is greater than the magnitude of the preceding $(l-1)$ vectors and the subsequent $(l-1)$ vectors. For example, if $l = 5$ the 100th vector is longer than vectors 96 to 99 and longer than vectors 101 to 104. An appropriate average to represent the focal point in the pathway $P(\eta_l(s))$ at q (for a specified l) can be conveniently defined as eta hat $\hat{\eta}_q(s)$;

$$\hat{\eta}_q(s) = \left(\eta_{l,a-1}(s) + \eta_{l,a}(s) + \eta_{l,b-1}(s) + \eta_{l,b}(s)\right)/4$$
with $a = l\lfloor m(q)/l \rfloor$ and $b = l\lceil m(q)/l \rceil$ and $q \in \mathbb{R}_l$. \tag{13}

The set R_l is defined below. Alternatively, the foci are implied by $Y_r(\gamma)$ as described in Section 3.

21. The focal point at the start of Q_q is $\hat{\eta}_{q+1}(s)$ and that at the end of Q_q is $\hat{\eta}_q(s)$. When $2l|q$ the $\vec{m}$ of Q_{q-1} and Q_{q+1} oscillate about $\hat{\eta}_q(s)$ such that the *distal* pseudo-spiral of a $P(\mathcal{R}_{r+1})$ and the *proximal* pseudo-spiral of its neighbouring $P(\mathcal{R}_r)$ allow the following equivalence

$$\hat{\eta}_{(q-1)}(s) \cong \hat{\eta}_q(s) \cong \hat{\eta}_{(q+1)}(s) \text{ when } 2l|q. \tag{14}$$

22. Convergence can be placed at $q = 1$ since $\eta_{l,\infty}(s) = \hat{\eta}_0(s) \cong \hat{\eta}_1(s)$ and pursuing $\eta_{l,n}(s)$ any further than $\hat{\eta}_1(s)$ merely adds computational burden without benefit, see Figure 5. We can now comfortably revert to $q \in \mathbb{N}_1$ as we no longer require $q = 0$.

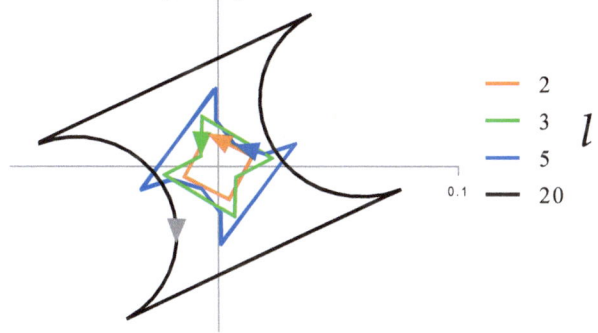

Figure 5. Vectors in $P(\eta_l(\rho_i))$ for the 3000th *zero* near $q = 1$ with the $\vec{m}$ for $m = m(1)$ indicated with an arrow for each of four values of l. For each l, the two l^{th} vectors shown are sufficient to reliably estimate $\eta_l(s)$ when $q = 1$, so that $\hat{\eta}_q(s) \cong \eta_l(s)$ when $q = 1$.

23. The relationship between an r and the q related to a pseudo-convergence depends on l and the geometry near each $\vec{m}$ when $m \in M_l$. The mapping of an r onto a q is captured in the set $R_l = \{q_1, q_2, q_3 \dots q_r \dots\}$. The set R_l is not to be confused with the set $\mathcal{R}_r$ being the $\vec{m}$ in a $P(\mathcal{R}_r)$. For any $l \geq 2$ we take odd values of q, and the even values of q which satisfy $2l|q$. We then remove the odd values of q which lie either side of the values of q for which $2l|q$. We could include $q = 0$, since this specifies convergence but are happy to accept $q_1 = 1$ in its place. We order this set as $q \in \mathbb{N}_1$ rises and index the elements by $r \in \mathbb{N}_1$ and so generate $R_l = \{q_1, q_2, q_3 \dots q_r \dots\}$. We state R_l formally using "\" to mean the set theoretic difference, thus

$$R_l = \{q_r\} = \{\{q: 2|(q+1)\}, \{q: 2l|q\}\} \setminus \{q: 2l|(q+1) \text{ or } 2l|(q-1)\}. \tag{15}$$

15

A few sets identifying the pseudo-convergences are;
$$R_2 = \{1, 4, 8, 12, 16, 20, 24, 28, 32 \ldots\},$$
$$R_3 = \{1, 3, 6, 9, 12, 15, 18, 21, 24 \ldots\},$$
$$R_4 = \{1, 3, 5, 8, 11, 13, 16, 19, 21, 24, 27, 29, 32, 35 \ldots\} \text{ and}$$
$$R_5 = \{1, 3, 5, 7, 10, 13, 15, 17, 20, 23, 25, 27, 30, 33 \ldots\}.$$

To incorporate Euler's *zeta* as a special case (rather than letting $l \to \infty$) we add that if $l = 1$ then $R_1 = \{1, 3, 5, 7 \ldots\} = \mathbb{O}$ and notice that the elements of R_l for $l \geq 2$ follow this pattern until the lth term. At this point we reflect that with $q = 1$ Equation 11 (page 14) gives $m(1) = \tau$ for Euler's *zeta*, which is appropriate. The collapse of structures either side of $2l|q$ is illustrated in Appendix A2 on page 103.

24. The *inflection point* in a $P(\mathcal{R}_r)$ is located with the rth element of the ordered set $\chi_l = \{q_{\bar{r}}\}$, which contains even values of q but for $l \geq 2$ excludes those where $2l|q$;

$$\chi_l = \{q_{\bar{r}}\} = \{q: 2|q, q > 0\} \setminus \{q: 2l|q\} \text{ for } l \geq 2,$$
and $\chi_l = \mathbb{E}_2$ for Euler's *zeta* when $l = 1$. \hfill (16)

The bar in $\bar{r}$, clarifies that the subscripted object relates to the *inflection point* in the $P(\mathcal{R}_r)$, though it has an existence beyond the $P(\mathcal{R}_r)$. The elements have a similar format to those of R_l;

$$\chi_l = \{q_{\bar{1}}, q_{\bar{2}}, q_{\bar{3}}, \ldots q_{\bar{r}}, \ldots\}. \hfill (17)$$

If $l = 2$ then $\chi_2 = \{2, 6, 10, 14, 18, 22, 26, 30, \ldots q_{\bar{r}}, \ldots\}$.
If $l = 3$ then $\chi_3 = \{2, 4, 8, 10, 14, 16, 20, 22, 26, \ldots q_{\bar{r}}, \ldots\}$.
If $l = 4$ then $\chi_4 = \{2, 4, 6, 10, 12, 14, 18, 20, 22, 26, \ldots q_{\bar{r}}, \ldots\}$,
and $\chi_5 = \{2, 4, 6, 8, 12, 14, 16, 18, 22, 24, 26, 28, 32 \ldots\}$ etc.

25. When $l \geq 2$ we can also define $q_{\bar{r}}$ inductively as $q_{\bar{1}} = 2$ and then $q_{\overline{r+1}} = q_{\bar{r}} + 2$ if $2l \nmid (q_{\bar{r}} + 2)$ and $q_{\overline{r+1}} = q_{\bar{r}} + 4$ if $2l|(q_{\bar{r}} + 2)$. The elements in the ordered set χ_l rise in increments of 2 until $l - 1$ terms have been reached, and so if $\kappa = l - 1$ we have

$$\left\{\frac{q_{\bar{r}}}{2}: \text{for } r < l\right\} = \{1, 2, 3, \ldots (l-1)\} = \{1, 2, 3, \ldots \kappa\}.$$

To allow application in Euler's *zeta* we have
$$\chi_1 = \{2, 4, 6, 8, 10 \ldots q_{\bar{r}} \ldots\} = \mathbb{E}_2.$$

26. In $\eta_l(s)$ each $P(\mathcal{R}_r)$ has a *principal-axis* which has an argument and a magnitude related to the vector $\vec{\mathcal{R}}_r$. The *principal-axes* have end points approximated by $\hat{\eta}_{q_{r+1}}(s)$ and $\hat{\eta}_{q_r}(s)$ with $q_r \in R_l$.

Alternatively, end points are those implied by $Y_r(u)$, see Section 3. We now have all the $\vec{m}$ in $\mathcal{R}_r$ with m in the specified interval using the floor function

$$\mathcal{R}_r = \left\{\vec{m} : \left\lfloor \left(e^{\frac{q_r\pi}{tl}} - 1\right)^{-1} \right\rfloor > m \geq \left\lfloor \left(e^{\frac{q_{r+1}\pi}{tl}} - 1\right)^{-1} \right\rfloor\right\}$$

$$r \in \mathbb{N}_1,\ q_r \in \mathbb{R}_l. \tag{18}$$

It is also acceptable to define the set using $q_{\tilde{r}} \in \chi_l$ as

$$\mathcal{R}_r = \left\{\vec{m} : \left\lfloor \left(e^{\frac{(q_{\tilde{r}}-1)\pi}{tl}} - 1\right)^{-1} \right\rfloor > m \geq \left\lfloor \left(e^{\frac{(q_{\tilde{r}}+1)\pi}{tl}} - 1\right)^{-1} \right\rfloor\right\}. \tag{19}$$

For large t the set $\mathcal{R}_r = \{\vec{m} : \lfloor tl/((q_{\tilde{r}}-1)\pi)\rfloor > m \geq \lfloor tl/((q_{\tilde{r}}+1)\pi)\rfloor\}$. The *principal-axis* of a $P(\mathcal{R}_r)$ preserves its orientation under changes in σ (see below). The final and largest structure in the pathway is $P(\mathcal{R}_1)$, whose *principal-axis* has magnitude $\sqrt{l}$ when $\sigma = 1/2$.

27. The *principal-axis* of a $P(\mathcal{R}_r)$ has a magnitude $|\vec{\mathcal{R}}_r|$ which we approximate to $|\Delta\hat{\eta}_{q_r}|$ and to the separation of the convergences of $Y_r(u)$, (see below). The magnitude $|\Delta\hat{\eta}_{q_r}|$ is

$$|\Delta\hat{\eta}_{q_r}| = |\hat{\eta}_{q_r}(s) - \hat{\eta}_{q_{r+1}}(s)|. \tag{20}$$

Appendix A3 (page 103) tabulates some ratios of $|\Delta\hat{\eta}_{q_r}|/|\Delta\hat{\eta}_{q_1}|$ for $l = 2$, with Appendix A4 (page 105) tabulating calculations for $l = 5$ for different real domains, together with calculations for the differentials.

28. A real nu, valid for all r and $q_{\tilde{r}}$ rising beyond the existence of the $P(\mathcal{R}_r)$, is defined as

$$v_r = \frac{1}{2}\left(\left(e^{\frac{q_{\tilde{r}}\pi}{tl}} - 1\right)^{-1} + \left(1 - e^{-\frac{q_{\tilde{r}}\pi}{tl}}\right)^{-1}\right) \text{ for all } l \text{ and all } t. \tag{21}$$

In each $P(\mathcal{R}_r)$ a $\vec{m}$ designated $\vec{v}_r$ lies near the *inflection point* with $m = [v_r]$. For large t, that m is $m = [tl/(q_{\tilde{r}}\pi)]$ and $v_1 \approx tl/2\pi$. The following are helpful; $v_r = v_1 2/q_{\tilde{r}}$ and if $l > \kappa$ then for $r < \kappa$ we have $v_r = v_1/r$. In Euler's zeta $v_1 \approx t/2\pi$ consistent with $l = 1$. Using $t = 740$ (not a *zero*) Figure 6 (overleaf on page 18) shows the $\vec{m}$, for which $l|m$, following the progress of a $P(\mathcal{R}_1)$ and having magnitudes $(l-1)m^{-\sigma}$. There are gaps between *neighbour*ing vectors for which $l|m$ which are filled by near-regular polygons (squares) with one missing side.

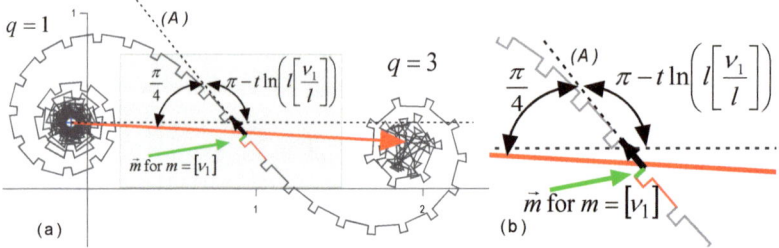

Figure 6. (a) Paired pseudo-spirals of $P(\mathcal{R}_1)$ for $t = 740$, $\sigma = 1/2$, $l = 4$ from $q = 3$ at $m = 314$ to $m = 942$ at $q = 1$. The red arrow is $\vec{\mathcal{R}}_1$ with length $2 = \sqrt{4}$. (b) Detail from (a).

In Figure 6, $\vec{m}$ with $m = \left[\left(e^{q_{\bar{r}}\pi/(tl)} - 1\right)^{-1}\right]$ for $q_{\bar{r}} = 2$ is shown in green at $m = 471$ and sits near the *inflection point*. Such a vector may point in an unhelpful direction unless $l|m$, the nearest having $m = l[v_r/l]$ is shown in black at $m = 472$ and its argument is a good estimate for $\arg(A)$, representing the tangent to a smooth curve $Y_r(u)$, which we imagine following the $P(\mathcal{R}_r)$. That argument is π $t\ln(l[v_r/l])$. The tangent crosses the *principal-axis* at an acute angle close to $\pi/4$. The vector $\vec{\mathcal{R}}_r$ in red, with magnitude $\sqrt{4}$, has an orientation displaced by π from the sum of the two illustrated arguments and is $\arg(\vec{\mathcal{R}}_1) = \pi/4 - t\ln\left(l\left[\frac{v_1}{l}\right]\right) = \pi/4 - t\ln(l[t/2\pi])$.

29. In summary, v_r locates the *inflection point* on $Y_r(u)$. An $\vec{m}$ is called $\vec{v}_r$ for which the associated $m = [v_r]$ and if $l \nmid [v_r]$ then that $\vec{v}_r$ lies in the near-regular open polygon or star polygon in $P(\eta_l(s))$ and its relationship to the $\arg(\vec{\mathcal{R}}_r)$ would be intricate. A tangential vector located nearby at $m = l[v_r/l]$ has $\arg(\vec{\mathcal{R}}_r) = \pi/4 - t\ln(l[v_r/l])$; noting that the addition of π for when $l|m$, cancels with the π associated with $\vec{\mathcal{R}}_r$ being directed from *distal* to *proximal*.

30. Nu separates the *proximal* and *distal* vectors within a $P(\mathcal{R}_r)$. In $\mathcal{R}_1$ there are about 3 times as many $\vec{m}$ in the *distal* as in the *proximal* part since the *proximal* has $\frac{tl}{3\pi} < m < \frac{tl}{2\pi}$ and the *distal* part has $\frac{tl}{2\pi} < m < \frac{tl}{\pi}$. In sequence, the ratios for the number of vectors in the *distal* to *proximal* parts of $P(\mathcal{R}_r)$ will be very close to $\frac{3}{1}, \frac{5}{3}, \frac{7}{5}, \frac{9}{7} \ldots \frac{2r+1}{2r-1}$, with the approximation being best with large t and small r.

31. A smooth curve $Y_r(u)$ closely follows each $P(\mathcal{R}_r)$ with an *inflection point* superimposed on the vector $\vec{v}_r$ where progression changes from clockwise to anticlockwise. The *proximal* clockwise spiral $Y_{r,p}(u)$ has

$u \leq v_r$ and the *distal* anticlockwise spiral $Y_{r,d}$ has $v_r < u$. The tangent at the *inflection point* $u = v_r$ bisects the *principal-axis* when $\sigma = 0$, crossing it at an angle of $3\pi/4$.

32. The paired spirals encoded in $Y_r(u)$, with $Y_{r,p}(u)$ and $Y_{r,d}(u)$, describe *proximal* and *distal* congruent spirals when $\sigma = 0$ that are also congruent with Euler's spiral. When $\sigma \neq 0$ $Y_{r,p}(u)$ and $Y_{r,d}(u)$ are no longer congruent but an average of the *proximal* and *distal* spirals, both placed in Quadrant I, is $\frac{1}{2}\left(Y_{r,p}(u) - Y_{r,d}(u)\right)$ which behaves as a single Euler's spiral when $0 \leq \sigma \leq 1$ and t is large.

33. A vector $\vec{\mathcal{R}}_r$, representing the *principal-axis* of a $P(\mathcal{R}_r)$ in $P(\zeta_n(s))$ is an object in its own right. It has a magnitude which we prove later, but state here to be

$$|\vec{\mathcal{R}}_r| = \sqrt{l}\, v_1^{(1/2-\sigma)} r^{(\sigma-1)} \tag{22}$$

with an argument

$$\arg(\vec{\mathcal{R}}_r) = \frac{5\pi}{4} + t \ln\left(\frac{r}{[v_1]}\right) \text{ for } l = 1. \tag{23}$$

A vector $\vec{\mathcal{R}}_r$, representing the *principal-axis* of a $P(\mathcal{R}_r)$ in $P(\eta_l(s))$ has magnitude

$$|\vec{\mathcal{R}}_r| = \sqrt{l}\, v_1^{(1/2-\sigma)} \left(\frac{q_{\bar{r}}}{2}\right)^{(\sigma-1)} \text{ with } q_{\bar{r}} \in \chi_l, \tag{24}$$

simplifying to

$$|\vec{\mathcal{R}}_r| = \sqrt{l/r}\, v_r^{(1/2-\sigma)} \text{ when } l > r,$$
$$\text{or } |\vec{\mathcal{R}}_r| = \sqrt{l}\, v_1^{(1/2-\sigma)} r^{(\sigma-1)}, \tag{25}$$

as in Equation 22. After taking the nearest integer $[v_1/l]$, the argument of an $\vec{\mathcal{R}}_r$ is

$$\arg(\vec{\mathcal{R}}_r) = \frac{\pi}{4} - t\ln\left(\frac{2l}{q_{\bar{r}}}\left[\frac{v_1}{l}\right]\right), \tag{26}$$

which simplifies to

$$\arg(\vec{\mathcal{R}}_r) = \frac{\pi}{4} - t\ln\left(\frac{l}{r}\left[\frac{v_1}{l}\right]\right) \text{ when } r < l. \tag{27}$$

In addition, since $v_1 \approx tl/(2\pi)$ we can see that

$$\arg(\vec{\mathcal{R}}_r) = \frac{\pi}{4} + t\ln\left(\frac{q_{\bar{r}}}{2l}\right) - t\ln\left(\left[\frac{t}{2\pi}\right]\right). \tag{28}$$

which simplifies to

$$\arg(\vec{\mathcal{R}}_r) = \frac{\pi}{4} + t\ln\left(\frac{r}{l}\right) - t\ln\left(\left[\frac{t}{2\pi}\right]\right) \text{ when } r < l. \tag{29}$$

34. For high values of t the following are useful reformulations;
$$|\vec{\mathcal{R}}_r| = \sqrt{l}\,(tl/(2\pi))^{(1/2-\sigma)} r^{(\sigma-1)},$$
$$|\vec{\mathcal{R}}_r| = l^{(1-\sigma)}(t/(2\pi))^{(1/2-\sigma)} r^{(\sigma-1)},$$
$$|\vec{\mathcal{R}}_r| = (l/r)^{(1-\sigma)}(t/(2\pi))^{(1/2-\sigma)},$$
and for all $r < l$, when

$\sigma = 1$ the $|\vec{\mathcal{R}}_r| = \sqrt{l/v_1} \approx \sqrt{2\pi/t}$,
$\sigma = 0$ the $|\vec{\mathcal{R}}_r| = \sqrt{lv_1}/r \approx l/r\sqrt{t/(2\pi)}$,
$\sigma = 1/2$ the $|\vec{\mathcal{R}}_r| = \sqrt{l/r}$.

35. The function $\bar{h}_{l,r}(s)$ is a partial series to r terms $\bar{h}_{l,r}(s) = \sum_{r=1}^{r} \vec{\mathcal{R}}_r$ driven by $q_{\bar{r}} \in \chi_l$. The $\vec{\mathcal{R}}_r$ shares its magnitude and argument with the *principal-axis* of a $P(\mathcal{R}_r)$ when $r < \kappa$, but exists when $r > \kappa$ without a $P(\mathcal{R}_r)$. Because $q_{\bar{r}}$, has the subscript "r bar", and this distinguishes the vector, the *bar* is re-used in the function $\bar{h}_{l,r}(s)$. The series $h_{l,\tau}(s)$ for all $l \geq 2$, has a relationship with $\eta_l(s)$, and is defined as;

$$\bar{h}_{l,\tau}(s) = \sum_{r=1}^{\tau} \vec{\mathcal{R}}_r \text{ with } \vec{\mathcal{R}}_r \text{ following } q_{\bar{r}} \in \chi_l. \tag{30}$$

At a *zero*, the pathway of $P\left(\bar{h}_{l,\tau}(s)\right)$ after its *kappa* vector $\vec{\mathcal{R}}_\kappa$ has a relationship with the *proximal* part of $P(\eta_l(s))$ in which $m < \kappa l$. The final pseudo-convergence can be refined from the point $\bar{h}_{l,\tau}(s)$ to an average of points near $\bar{h}_{l,\tau}(s)$ using $\hat{h}(s)$;

$$\hat{h}(s) = \left(\bar{h}_{l,a-1}(s) + \bar{h}_{l,a}(s) + \bar{h}_{l,b-1}(s) + \bar{h}_{l,b}(s)\right)/4$$
with $a = l\lfloor\tau/l\rfloor$ and $b = l\lceil\tau/l\rceil$. \hfill (31)

$P(\bar{h}_{l,\tau}(\rho))$ runs counter to $P(\eta_l(\rho))$, and what is *proximal* for one pathway is *distal* for the other.

36. The separation of *proximal* and *distal* parts of $P(\zeta_n(s))$ and of $P(\eta_l(s))$ for $l > \kappa$ occur when $|\vec{\mathcal{R}}_r| \cong |\vec{m}|$ which is when $|\vec{\mathcal{R}}_r| = l^{(1-\sigma)}\left(\frac{t}{2\pi}\right)^{(-\sigma/2)} = l(\kappa l)^{-\sigma}$. The vectors of $P(\eta_{l,n}(s))$ and $P(h_{l,r}(s))$ get smaller as m and r rise, but since they are complementary pathways at *zeros*, there must be a crossover-point where $|\vec{\mathcal{R}}_r| \cong |\vec{m}|$ and this is at the κl term.

37. The function $h_{l,r}(s)$ is a partial series to r terms driven by r that is simpler than $\bar{h}_{l,r}(s)$;

$$h_{l,r}(s) = \sum_{r=1}^{r} \vec{\mathcal{R}}_r \text{ with } \vec{\mathcal{R}}_r \text{ following } q_{\vec{r}} \in \mathbb{E}_2. \tag{32}$$

The series $h_{l,\kappa}(s)$ to *kappa* terms is applicable to Euler's *zeta* and if $\kappa < l$ it applies to $\eta_l(s)$;

$$h_{l,\kappa}(s) = \sum_{r=1}^{\kappa} \vec{\mathcal{R}}_r \text{ with } \vec{\mathcal{R}}_r \text{ following } q_{\vec{r}} \in \mathbb{E}_2 \text{ if } \kappa < l.$$

38. The partial series $\eta_{l,\kappa l}(s)$ to κl terms is specifically

$$\eta_{l,\kappa l}(s) = \zeta_{\kappa l}(s) - l \sum_{n=1}^{\kappa} (ln)^{-s} \tag{33}$$

which can be restated as

$$\eta_{l,\kappa l}(s) = \zeta_{kl}(s) - l^{(1-s)} \zeta_\kappa(s). \tag{34}$$

When $l > \kappa$ the symmetries in $P(\eta_l(s))$ are overt, and when $l \approx [\tau/\kappa]$ and $s \in \{\rho_i\}$ the term $\zeta_{\kappa l}(s) \approx 0$. It is argued that when $l \approx [\tau/\kappa]$ and $s \in \{\rho_u\}$ then the term $\zeta_{\kappa l}(s) \approx 0$. However, if there were a ρ_u for which $\zeta_{\kappa l}(s) \not\approx 0$ the arguments below show that this is not an anxiety.

39. A finite series of vectors indexed by $\mathfrak{n}$ (a script) and summated to $\mathfrak{n}$ terms (a Fraktur) is

$$\ell_{l,\mathfrak{n}}(s) = -l^{(1-s)} \sum_{\mathfrak{n}=1}^{\mathfrak{n}} \mathfrak{n}^{-s} = \sum_{\mathfrak{n}=1}^{\mathfrak{n}} \vec{\mathcal{L}}_\mathfrak{n} \tag{35}$$

with $\vec{\mathcal{L}}_\mathfrak{n} = -l^{(1-s)} \mathfrak{n}^{-s}$. $\tag{36}$

The notation $\mathfrak{n}$ for the final term is analogous to r and n for the other series. However, $\ell_{l,\mathfrak{n}}(s)$ is principally needed to *kappa* terms and so $\ell_{l,\kappa}(s)$ is;

$$\ell_{l,\kappa}(s) = -l^{(1-s)} \sum_{\mathfrak{n}=1}^{\kappa} \mathfrak{n}^{-s} = \sum_{\mathfrak{n}=1}^{\kappa} \vec{\mathcal{L}}_\mathfrak{n}.$$

If $\mathfrak{n}$ is allowed to rise to *tau* then $\ell_{l,\tau}(s)$ has a final pseudo-convergence before diverging, with τ defined as for Euler's *zeta*, Equation 9 on page 11.

40. *Psi* is the line of reflection for the arguments for all paired vectors and the line of reflection for the difference between all neighbouring vectors. It is the line of reflection for the paired pathways $P(h_{l,\kappa}(s))$ and $P(\ell_{l,\kappa}(s))$, or for $P(h_{1,\kappa}(s))$ and $P(\zeta_\kappa(s))$, when $\sigma = 1/2$ and those pathways set of from the origin. This follows from the paired vectors having equal magnitudes when $\sigma = 1/2$. The first vector of $\ell_{l,\kappa}(s)$, $\vec{L}_1$ has an orientation of $-(t \ln(l) + \pi)$ and the orientation of the first vector of $h_{l,\kappa}(s)$ is $\pi/4 - t\ln(l[v_1/l])$. The argument ψ_0 is half the sum of the two arguments. The line of reflection $\vec{\psi}$, is made of two rays $\vec{\psi}_0$ and $\vec{\psi}_\pi$ of unspecified magnitude and having arguments ψ_0 and ψ_π. The argument ψ_0 when $l \geq 2$ in $\eta_l(s)$ is

$$\psi_0 = \arg(\vec{\psi}_0) = -\frac{t}{2}\ln\left(l^2 \begin{bmatrix} v_1 \\ l \end{bmatrix}\right) - \frac{3\pi}{8}$$
$$= -t\ln(l) - \frac{t}{2}\ln\left(\begin{bmatrix} t \\ 2\pi \end{bmatrix}\right) - \frac{3\pi}{8}, \tag{37}$$

and if $l = 1$, for Euler's zeta

$$\psi_0 = \arg(\vec{\psi}_0) = \frac{5\pi}{8} - \frac{t}{2}\ln\left(\begin{bmatrix} t \\ 2\pi \end{bmatrix}\right). \tag{38}$$

The argument ψ_π is displaced clockwise from ψ_0 and so $\psi_\pi = \psi_0 - \pi$. This *Psi* is not to be confused with ψ used elsewhere such as for the second Chebyshev function. The *kappa* vectors cross $\vec{\psi}$ when $s \in \{\rho\}$ and at that crossing on $\vec{\psi}$, a line normal to $\vec{\psi}$ can be imagined, we call this a *"projection of the critical line"* (not shown till Figure 40 page 67).

2.5. Axiom 2, Axiom 3 and approximations

Axiom 2 is that $\Delta\vartheta$, the difference between the arguments of vectors separated by $l-1$ intervening vectors in $P(\eta_l(s))$, is a function of t and m;

$$\Delta\vartheta = t\ln\left(\frac{m+l}{m}\right) \text{ with } 0 < \Delta\vartheta < \infty. \tag{39}$$

Equation 39 applies for all l. In reduced form, $\Delta\vartheta$ is presented as $\Delta\theta$ in the range $-\pi < \Delta\theta < \pi$ as

$$\Delta\theta = q_{\bar{r}}\pi - t\ln\left(\frac{m+l}{m}\right) \text{ with } q_{\bar{r}} \in \chi_l, \tag{40}$$

which in appropriate circumstances approximates to

$$\Delta\theta = q_{\bar{r}}\pi - tl/m. \tag{41}$$

The r of the $P(\mathcal{R}_r)$ in $P(\eta_l(s))$ is a count of the number of paired-spirals from convergence and for all $l \geq 2$ it is recoverable as

$$r = q_{\bar{r}}/2 - \lfloor q_{\bar{r}}/(2l) \rfloor. \tag{42}$$

Axiom 2, captured in Equations 39–42, is a prescription of pathway shape as a function of t and l.

Axiom 3 is that the magnitudes of neighbouring vectors differ from one another as a function of σ and m and as a consequence change in pathway shape is a differential shrinkage determined by

$$\frac{\partial}{\partial \sigma} m^{-\sigma} = -\ln(m)\, m^{-\sigma}. \tag{43}$$

These two *Axioms* and the Functional Equation (*Axiom 1*) are the only rules dictating the behavior of Euler's *zeta*, $\eta_l(s)$ and related series. The naïve premise is that these alone are sufficient to prove or disprove RH; with the "*if not, why not*" voicing the sober premise.

Important transitions in a pathway are consequent upon *Axiom 2*. To $m \in \mathbb{N}_1$ and $\Delta \vartheta \in \mathbb{R}$, for this section, we add $x \in \mathbb{N}_1$ and $\mu, \mu_1, \mu_2 \in \mathbb{R}$. Significant changes in a pathway take place either side of some $\vec{m}$ associated with an $m = [\mu]$, occurring at a specific $\Delta \vartheta = x\pi$, and for which $x = (t/\pi)\ln((\mu_1 + 1)/\mu_1)$ giving $\mu_1 = (e^{x\pi/t} - 1)^{-1}$ with $\mu_1 < \mu$, and for $x = (t/\pi)\ln((\mu_2 - 1)/\mu_2)$ giving $\mu_2 = (1 - e^{-x\pi/t})^{-1}$ with $\mu < \mu_2$. We are then able to choose $\mu = (\mu_1 + \mu_2)/2$ as

$$\mu(x, t) = \frac{1}{2}\left((e^{x\pi/t} - 1)^{-1} + (1 - e^{-x\pi/t})^{-1}\right) \tag{44}$$

from which m, if required, is recovered as $m = [\mu]$. The precision of Equation 44 is important at low values of t, but at high values we substitute $y = x\pi/t$ and use $e^y = \sum_{n=0}^{\infty} y^n/n!$ to give $(e^y - 1) = y + y^2/2 + y^3/6 + + y^4/24 \ldots$ and $(1 - e^{-y}) = y - y^2/2 + y^3/6 - y^4/24 \ldots$. This allows us to determine $((e^y - 1)^{-1} + (1 - e^{-y})^{-1})/2$ with

$$\mu(y) = \frac{1}{2}\left(\left(y + \frac{y^2}{2} + \frac{y^3}{6} + \frac{y^4}{24} + \frac{y^5}{120} + \cdots\right)^{-1} \right.$$
$$\left. + \left(y - \frac{y^2}{2} + \frac{y^3}{6} - \frac{y^4}{24} + \frac{y^5}{120} - \cdots\right)^{-1}\right). \tag{45}$$

Then given $\frac{1}{a} + \frac{1}{b} = \frac{b+a}{ab}$, Equation 45 yields the interesting series

$$\mu(y) = y\left(y^2 + \frac{y^4}{12} + \frac{y^6}{360} + \frac{y^8}{20160} + \frac{y^{10}}{1814400} + \frac{y^{12}}{239500800} + \frac{y^{14}}{43589145600} \cdots\right)^{-1}. \tag{46}$$

If $t \gg x\pi$, which it is if we are interested in RH, then $y \ll 1$ and we accept $\mu(x,t) = y/y^2 = y^{-1} = t/(x\pi)$. In passing, and of no importance to the arguments of this monograph, it is noted that Equation 46 equates to

$$\mu(y) = y\left(2\sum_{n=1}^{\infty} \frac{y^{2n}}{(2n)!}\right)^{-1} = \frac{y}{2\cosh(y) - 2}. \tag{47}$$

2.6. Pathway symmetry about a line of reflection

In a partial Euler's zeta $\zeta_\tau(s)$, matched $\vec{m}$ from $P(\zeta_\kappa(s))$ and $\vec{\mathcal{R}}_r$ from $P(h_{1,\kappa}(s))$ have mirror image arguments about a common line of reflection but their magnitudes, which equate when $\sigma = 1/2$, differ when $\sigma \neq 1/2$. If one knew nothing of analytical continuation and nothing about RH this simple observation suggests both a population of zeros and limits these zeros to having $\sigma = 1/2$.

In an $\eta_l(s)$, if $l > \kappa$ then paired vectors $\vec{\mathcal{L}}_n$ from $\ell_{l,n}(s)$, and $\vec{\mathcal{R}}_r$ from $h_{l,r}(s)$ have mirror image arguments as far as the kappa vectors $\vec{\mathcal{L}}_\kappa$ and $\vec{\mathcal{R}}_\kappa$. Once more, magnitudes which equate when $\sigma = 1/2$ differ when $\sigma \neq 1/2$. This underlies a proof of RH. If $l > \kappa$ and the intersection of the kappa vectors $\vec{\mathcal{L}}_\kappa$ and $\vec{\mathcal{R}}_\kappa$ is accommodated in $x = f(\dot{\kappa})$ then

$$\eta_l(s) \approx \zeta_{\kappa l}(s) + \ell_{l,\kappa-1}(s) + x\vec{\mathcal{L}}_\kappa - x\vec{\mathcal{R}}_\kappa - h_{l,\kappa-1}(s). \tag{48}$$

The symmetries between $P(\ell_{l,\kappa}(s))$ and $P(h_{l,\kappa}(s))$ are striking, though $\zeta_{\kappa l}(s)$ introduces a small irritation. Fortunately, near the zeros $\zeta_{\kappa l}(s)$ can be ignored so exposing the symmetries. This is best satisfied with $l = [\tau/\kappa]$ when $\zeta_\tau(\rho) \approx 0$. Equation 48, derived from the geometry, resembles the famous Riemann-Siegel formula which has two finite Dirichlet series to $[(t/(2\pi))^{1/2}]$ terms, forming an approximate Functional Equation with an asymptotic error term.

2.7. The function lambda

A function $\lambda_l(s,x)$, which shares zeros with $\zeta(s)$, is essentially $\eta_l(s)$ agnostic of $\zeta_{\kappa l}(s)$ with $x = f(\dot{\kappa})$ specifying the intersection of the kappa vectors;

$$\vec{\lambda}_l(s,x) = \vec{\ell}_{l,\kappa-1}(s) + x\vec{\mathcal{L}}_\kappa - x\vec{\mathcal{R}}_\kappa - \vec{h}_{l,\kappa-1}(s). \tag{49}$$

We can also think of $\vec{\lambda}_l(s,x)$ joining the ends of the pathways in $\mathbb{R}^2$ with the pathways anchored at the origin

$$P(h_{l,\kappa-1}(s) + x\vec{\mathcal{R}}_\kappa) + \vec{\lambda}_l(s,x) = P(\ell_{l,\kappa-1}(s) + x\vec{\mathcal{L}}_\kappa).$$

This helps because $P(\ell_{l,\kappa-1}(s) + x\vec{\mathcal{L}}_\kappa)$ and $P(h_{l,\kappa-1}(s) + x\vec{\mathcal{R}}_\kappa)$ are mirror images when $\mathrm{Re}(s) = 1/2$ and they meet when $t \in \{t_i\}$, otherwise they do not meet. When $\sigma \neq 1/2$ the symmetry breaks and the pathways cannot meet. Symmetry breaking limits the *zeros* to the *critical line* irrespective of the value of t. It immediately follows that the pathways for the zeros of the derivative $\lambda'_l(s,x)$ are limited to the right of the *critical line*. The derivatives magnify the pathways in different ways and reverse them in $\mathbb{R}^2$ such that symmetry can only be restored to create a zero for the differential when $\sigma > 1/2$. The differential is $\eta'_l = \lambda'_l + \zeta'_{\kappa l}$ and since $\zeta'_{\kappa l}(\rho) \neq 0$ comment is required. Nonetheless, the direction of restoration of symmetry is in agreement with Speiser's corollary [8].

2.8. The cardinality of the sets of proximal $\vec{m}$ and distal vectors $\vec{\mathcal{R}}_r$ in $\zeta_\infty(s)$ equate

In Euler's *zeta* the cardinality of the set of *proximal* vectors equates to the cardinality of the set $\{P(\mathcal{R}_r)\}$. Remembering that $\mathcal{R}_r$ is a set of $\vec{m}$ we state **Lemma (2)**;

$$\#\{\vec{m} : m \leq \kappa\} = \#\{\vec{\mathcal{R}}_r : \mathcal{R}_r \ni \vec{m} \text{ for which all } m > \kappa\} = \kappa. \tag{50}$$

Proof: of Equation 50 *Lemma (2)*. The *proximal* pathway has $\#\{\vec{m} : m \leq \kappa\} = \kappa$ and $\kappa = \left\lfloor \sqrt{t/(2\pi)} \right\rfloor$. In the *distal* pathway, as m rises, each $P(\mathcal{R}_r)$ has an *inflection point* at an m which minimises $|\varepsilon|$ in $t\ln((m \pm 1)/m) = 2\pi r \pm \varepsilon$ and each $P(\mathcal{R}_r)$ starts and ends at a change in $[t\ln((m+1)/m)/(2\pi)]$. Consequently at any $m = m'$ the number of $P(\mathcal{R}_r)$ that can be counted as r falls to 1 is $y = \#\{\vec{\mathcal{R}}_r : \mathcal{R}_r \ni \vec{m} \text{ for which all } m > m'\}$ then $t\ln((m'+1)/m') \approx 2\pi y$ and so if $m' = \kappa$ we have $t\ln((\kappa+1)/\kappa) \approx 2\pi y$, and $\ln(1 + 1/\kappa) \approx 2\pi y/t$ allowing the approximation $1/\kappa \approx 2\pi y/t$ into which we substitute $\kappa^{-2} \approx 2\pi/t$. Finally, turning to the integers we have $y = \kappa$.

2.9. Understanding pathway change as a consequence of Axiom 3

When $\eta_l(s)$ is plotted as $\eta_l(\sigma)$ for fixed t or as $\eta_l(t)$ for fixed σ we can imagine what is required for RH to fail. Figure 7 (overleaf on page 26) shows hypothetical curves for a ρ_u with a t_u, σ_α and σ_β. For a ρ_u that disproves RH there has to be a loop in $\eta_l(\sigma)$ for t_u which necessarily heads off to $1 + 0i$ as σ rises; something like the black loop in Figure 7. Crossing this loop at right angles, as prescribed by the partial differentials

$\frac{\partial \eta_l(s)}{\partial \sigma} = -i \frac{\partial \eta_l(s)}{\partial t}$, and passing in almost opposing directions are the two curves $\eta_l(t)$ for fixed σ_α and σ_β. The red curve is the true cycloid-like curve in $\eta_l(t)$ for fixed σ_c, which meets a true cycloid-like curve $\eta_l(\sigma)$ for fixed t_c (not shown). The cycloids meet at $\eta_l(\sigma_c + it_c)$, in the red square where $\eta'_l(\sigma_c + it_c) = 0$ with $\sigma_\beta - \sigma_c > \sigma_c - \sigma_\alpha$; the zero of the differential sits nearer the lower value of σ. This illustrates the challenge for the differential; if RH is false the *axioms* must be capable of generating a zero for the differential nearer to σ_α than σ_β and so at a value less than $1/2$, which is to the left of the *critical line*. This is a corollary of RH: if the differential cannot have a zero to the left of the *critical line* then RH has to be true. The corollary is Speiser's [8] but the arguments are different.

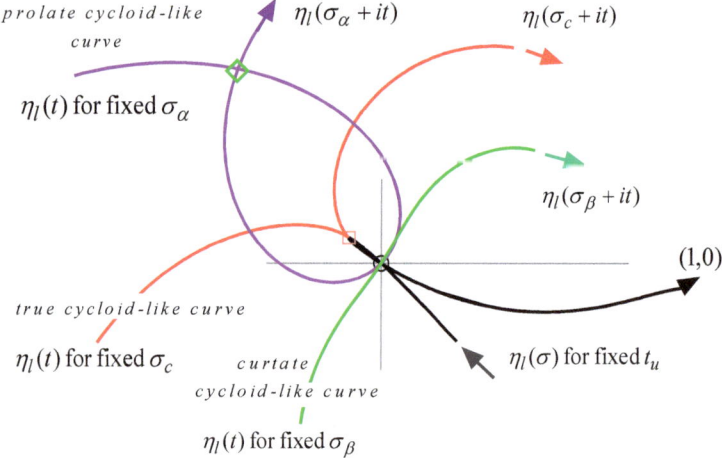

Figure 7. A curve in $\eta_l(\sigma)$ for fixed t_u in black, has a loop, and a double-point in the small black circle where $\eta_l(\sigma_\alpha + it_u) = \eta_l(\sigma_\beta + it_u) = 0$. A prolate cycloid-like curve in $\eta_l(t)$ for fixed σ_α in purple has a double-point $\eta_l(\sigma_\alpha + it_a) = \eta_l(\sigma_\alpha + it_b)$ in a green diamond. A curtate cycloid-like curve for $\eta_l(t)$ for fixed σ_β is shown in green. The true cycloid-like curve $\eta_l(\sigma_c + it)$ in red lies near the tip of the loop and has a zero differential at $\eta_l(\sigma_c + it_c)$ indicated with a red square.

This graphical context also helps us appreciate why it is impossible for RH to fail. If RH is to fail then a loop in $\eta_l(\sigma)$ must exist for all l. For an appropriate t, pathway change secondary to a rise in σ and $\frac{\partial}{\partial \sigma} m^{-\sigma} = -\ln(m) m^{-\sigma}$ (*Axiom 3*), can bring about a curve (without a double-point) in $\eta_l(\sigma)$ or under particular conditions, which are sensitive to l, a loop with a double-point in $\eta_l(\sigma)$. Curves capable of forming loops in $\eta_l(\sigma)$ as σ rises are solely a consequence of pathway change secondary to differential change in the magnitudes of the vectors with respect to distance along the pathway. In Figure 8(a) three pathways for P($\eta_l(s)$) are shown for σ = 1/2 in black, σ = 7/10 in red and σ =1 in blue, with the curve $\eta_l(\sigma)$ in green. The curve $\eta_l(\sigma)$ for t_{632} passes through zero when σ = 1/2 and then heads towards (1,0) for larger values of σ. There is no loop in $\eta_l(\sigma)$ in Figure 8(a), but the mechanisms of *Axiom 3* underlying the curve's retroversion are clear.

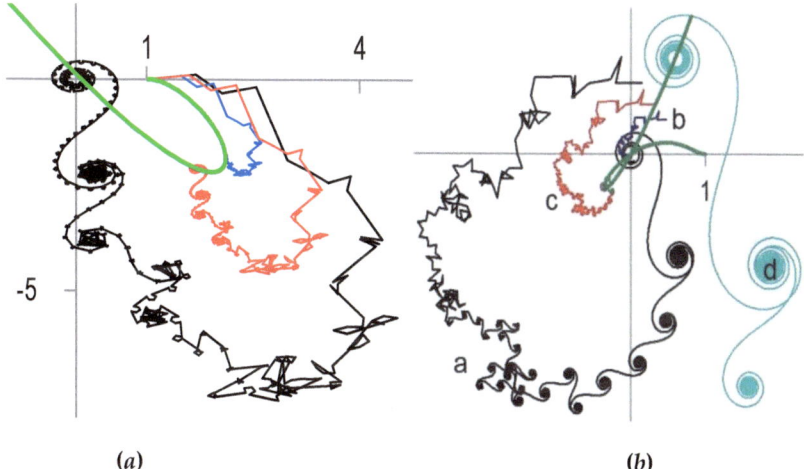

(a)   (b)

Figure 8. (a) P($\eta_3(\sigma + it_{632})$) for three values of σ showing the *proximo-distal shrinkage* of the pathway as σ rises. Matched vectors are parallel but *shrinkage* is greater at higher values of m as σ rises. A plot of $\eta_3(\sigma)$ in green is shown for fixed t. There is no loop or double-point but the mechanism underlying looping is evident. (b) A hypothetical loop for a ρ_u formed by the same *proximo-distal shrinkage*, see text for discussion.

Figure 8(b) shows a hypothetical loop in $\eta_l(\sigma)$ in green for a ρ_u. Pathway (a) in black is for $\rho_u = \sigma_\alpha + it_u$, pathway (b) in blue is for $\rho_u = \sigma_\beta + it_u$ and pathway (c) in red is for a value of σ satisfying $\sigma_\alpha < \sigma < \sigma_\beta$. Pathway (d) in light blue shows the final two paired pseudo-spirals for a value of $\sigma < \sigma_\alpha$. Figure 8(b) illustrates how the *proximo-distal shrinkage* of the pathway can create a loop as σ rises. Panels (a) and (b) in Figure 8 also illustrate the stability of the arg($\vec{\mathcal{R}}_r$). This stability, proved in Section 3, applies for all t and across the *critical strip*.

A further requirement, if RH were false, is that $\eta_l(\sigma)$ must be capable of forming a double-loop with a triple-point at zero for an $l = e^{2\pi k/t_u}$ for some $k \in \mathbb{N}_1$, (Roman k not Greek *kappa*). This is impossible since there are only two phenomena acting: (1) pathway layout and (2) the *proximo-distal* gradient which changes in a predictable way with Re(s) which at best can only allow a single loop.

3. The *Principal-Axis:* its magnitude and stability

The magnitude of the *principal-axis* of a P($\mathcal{R}_r$) and the stability of its argument are central to this work. We now prove **Lemma (3)** $\frac{\partial \arg(\vec{\mathcal{R}}_r)}{\partial \sigma} = 0$ and **Lemma (4)** $|\vec{\mathcal{R}}_r| = \sqrt{l/r}\, v_r^{1/2-\sigma}$ over the *critical strip* for high values of t. An equivalence to *Lemma (4)* is $|\vec{\mathcal{R}}_r| = \sqrt{l}\,(tl/2\pi)^{(1/2-\sigma)} r^{(\sigma-1)}$.

3.1. Paired Euler's spirals resemble a P($\mathcal{R}_r$)

This section uses i as an index for a summation with $c, s, \mu \in \mathbb{R}$ as *dummy* variables for integrals and $\rho \in \mathbb{R}$ for radius of curvature. These are not to be confused with s the domain of *zeta*, $\mu(x)$ the Mobius function, $i = \sqrt{-1}$ and ρ a *zero* of *zeta*. Euler spirals and Fresnel integrals are known for their applications in diffraction and in optimising curves in railway tracks and highways but are not widely known for their association with the *zeta* function.

Imagine paired Euler spirals in a Cartesian plane, one spiral labelled *proximal* in Quadrant I, and one labelled *distal* in Quadrant III. The arc of each spiral has a curvature inversely proportional to the arc length d, measured from the *inflection point* at the origin. The spirals are encoded in $E(d) \in \mathbb{C}$ with $d \to \infty$ for the *distal* spiral and $d \to -\infty$ for the *proximal* spiral; $E(d)$ can be parametised as $x_E(d) = \text{Re}(E(d))$ and $y_E(d) = \text{Im}(E(d))$;

$$E(d) = -\int_0^d \frac{s}{|s|} e^{is^2} ds = -\int_0^d \cos(s^2)\, ds - i \int_0^d \sin(s^2)\, ds$$

$$x(d) = -\int_0^d \cos(s^2)\, ds \text{ and } y(d) = -\int_0^d \sin(s^2)\, ds. \tag{51}$$

Notations are used with specific meanings. The formulation $E(d)$ describes the point $E(d)$ whilst the vector to that point is designated $\vec{E}(d)$. However, we allow $E(d)$ to imply the spiral generated whilst the integration follows the dummy variable s $\{s: 0 \le s \le d\}$. In contrast, $\vec{E}(d)$ is limited to being a vector and does not imply an encoded spiral. For $\vec{E}(d)$ the integration is solely a means to an end. Likewise, $Y_r(\gamma)$, to be described later, encodes paired spirals, whereas $\vec{H}(\sigma)$ solely represents its corresponding *principal-axis* and is understood solely as a vector.

The $E(d)$ spiral converges rapidly with convergences at $d = \pm\infty$ being predictable at low, but specific, magnitudes of d. Consequently, we define an approximation for the vector to $E(\infty)$, the point of convergence of the *distal* spiral, to be $\vec{E}_d = \vec{E}(\infty)$ with subscript d meaning *distal*;

$$\vec{E}_d \approx -\int_0^{d_1} \cos(s^2)\,ds - i\int_0^{d_2} \sin(s^2)\,ds$$

$d_1 = \sqrt{n\pi/2}$ and $d_2 = \sqrt{(n+1)\pi/2}$ with $n \in \mathbb{E}$ and $n > 2$. (52)

Likewise, $\vec{E}_p$ with p for *proximal* follows by negating d_1 and d_2 and is an approximation for the vector to $\vec{E}(-\infty)$, the point of convergence of the *proximal* spiral $E(-\infty)$ in Quadrant I. These approximations may seem superfluous but are used to help explain the stability and magnitude of the *principal-axis* of a $P(\mathcal{R}_r)$.

Paired Euler spirals, encoded in $\int_{-\infty}^{\infty} e^{is^2}\,ds$ have a *principal-axis* $\vec{E}_p - \vec{E}_d = 2\vec{E}_p$ running between the convergences and directed from *distal* to *proximal* forming an analogy with the paired spirals of a $P(\mathcal{R}_r)$ whose *principal-axis* equates to an $\vec{\mathcal{R}}_r$ and runs between focal points.

In the 1780's Euler proved $\text{Re}(\int_0^{\infty} e^{is^2}\,ds) = \text{Im}(\int_0^{\infty} e^{is^2}\,ds) = \sqrt{\pi/8}$ and so if we place the *principal-axis* of the paired Euler spirals on the line $x = y$ they have a magnitude of $\sqrt{\pi}$. Figure 9(a) (overleaf on page 30) compares a *distal* Euler spiral to $d = \sqrt{11\pi}$ (in red) and one scaled down by $\sqrt{2/\pi}$ (dashed-black) alongside the *distal* $P(\mathcal{R}_1)$ for a modest value of t when $l = 2$ and $\sigma = 1/2$ (in grey). The *distal* $P(\mathcal{R}_1)$ is appropriately rotated and translated.

Figure 9 shows, (1) that d need rise only a little before the location of convergence is easily determined, (2) at a modest value of t the $P(\mathcal{R}_r)$ is already nearly a smooth curve and resembles $E(d)$ and (3) at $\sigma = 1/2$ any deviation from similarity with $E(d)$ is small. The curve $E(d)$, plotted in Quadrant III, uses the negative Fresnel integrals, calculable as

$$x_E(d) = -d + \frac{d^5}{5 \times 2!} - \frac{d^9}{9 \times 4!} + \frac{d^{13}}{13 \times 6!} - \frac{d^{17}}{17 \times 8!} \cdots$$
$$= -\sum_{i=0}^{\infty} \frac{d^{4i+1}}{(4i+1)(2i)!} \quad (53)$$

$$y_E(d) = -\frac{d^3}{3} + \frac{d^7}{7 \times 3!} - \frac{d^{11}}{11 \times 5!} + \frac{d^{15}}{15 \times 7!} - \frac{d^{19}}{19 \times 9!} \cdots$$
$$= -\sum_{i=0}^{\infty} \frac{d^{4i+3}}{(4i+3)(2i+1)!}. \quad (54)$$

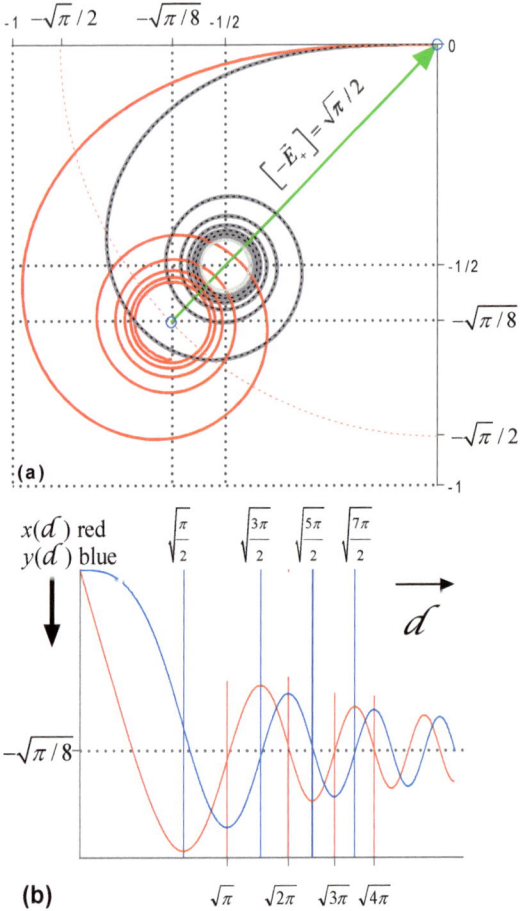

Figure 9. (a) A *distal* Euler spiral in red with $E(d) = -\int_0^d e^{is^2} ds$ and $\{d : 0 \leq d \leq \sqrt{11\pi}\}$. In grey the *distal* half of a $P(\mathcal{R}_1)$ using $l = 2$, $t_i = 55{,}271.699$, $\sigma = 1/2$ and in dashed black $\sqrt{l/(r\pi)}\, v_r^{(1/2-\sigma)} E(d) = \sqrt{2/\pi}\, E(d)$. The green vector represents $-\vec{E}_d$, the *distal* half of the *principal-axis* whose magnitude is $|\int_0^\infty e^{is^2} ds| = \sqrt{\pi}/2$ and argument $\pi/4$. (b) Low values of d permit excellent estimates of the location of the final convergence.

3.2. A smooth curve $Y_r(u)$ follows the vectors of a $P(\mathcal{R}_r)$ in an $\eta_l(s)$

A smooth curve $Y_r(u) = (x_r(u), y_r(u))$ is placed in $\mathbb{R}^2$ following a $P(\mathcal{R}_r)$; these objects share the subscript r which counts the $P(\mathcal{R}_r)$ from convergence. The curve $Y_r(u)$ follows the $\vec{m}$ where $l|m$ in $\eta_l(s)$ and falls upon all $\vec{m}$ in $\eta_2(s)$. The radius of curvature for $Y_r(u)$ is defined at $u \in \mathbb{R}$ to be $\rho_{Y_r}(u)$ and we equate this to the radius $\rho_{\mathcal{R}_r}(m)$ of a $P(\mathcal{R}_r)$ defined at each $m \in \{m \in \mathbb{N}_1 : l|m, \left\lceil \frac{tl}{(q_{\tilde{r}}+1)\pi} \right\rceil < m < \left\lceil \frac{tl}{(q_{\tilde{r}}-1)\pi} \right\rceil \}$. It is acceptable, to use the limits of $(q_{\tilde{r}} + 1)$ and $(q_{\tilde{r}} - 1)$ rather than (q_r) and $(q_r - 1)$ because when $q_r|2l$ we have $(q_r - 1), q_r$ and $(q_r + 1)$ all collocated with local oscillations making no advancement in $\mathbb{R}^2$. We let $v_r = tl/(q_{\tilde{r}}\pi)$ and if $m = [v_r]$ then the associated $\vec{m}$, designated $\vec{v}_r$, lies at the *inflection point*. We assume (for now) that the orientation of $\vec{v}_r$ is tangential to $Y_r(v_r)$ and is directed along the negative *x axis*, see Figure 10.

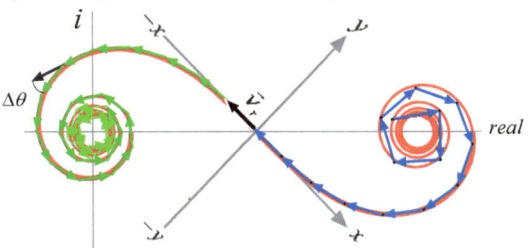

Figure 10. The vectors of a $P(\mathcal{R}_1)$ are shown in blue and green: $l = 2$, $t_i = 114.3202$ and $\sigma = 1/2$ have been chosen to place $\vec{\mathcal{R}}_1$ along the real axis in $\mathbb{C}$. A superimposed Cartesian (x, y) plane hosts the spirals of $Y_r(u)$ in red.

In Figure 10, *proximal* vectors are in blue in Quadrant I, and *distal* vectors are in green in Quadrant III. The difference between arguments of neighbouring $\vec{m}$ is $\Delta\theta = \pi - t/m$, and in general, the difference between arguments of vectors l apart is $\Delta\theta = q_{\tilde{r}}\pi - tl/m$.

We now determine the relationship between the radius of curvature and the arc length from the *inflection point* looking forwards as m rises and consider the general case when $l \geq 2$ and $q_{\tilde{r}} > 2$. Figure 11(a) shows a $P(\mathcal{R}_r)$ with radii drawn to the *proximal* ends of four tangential l^{th} vectors.

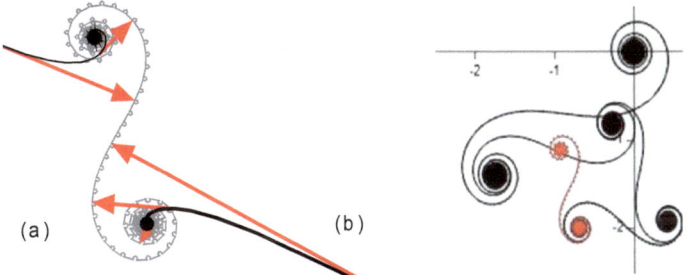

Figure 11. (a) A $P(\mathcal{R}_5)$ with $t_i = 55{,}271.699$ and $\sigma = 1/2$, from $m = 6767$ to 7997 with an *inflection point* at 7331, with $l = 5$ and $q_{\tilde{r}} = 12$. Example radii (red) are set at $t\ln(m) + \pi/2$ and drawn to the end of the $\vec{m}$ where $l|m$. Curves in black follow the centre of curvature as m rises. (b) The last 5 $P(\mathcal{R}_r)$ are shown with $P(\mathcal{R}_5)$ in red; $\chi_5 = \{2, 4, 6, 8, 12, \ldots q_{\tilde{r}}, \ldots\}$, $10 \notin \{\chi_5\}$.

The radius has a magnitude $|\rho_{\mathcal{R}_r}(m)|$, which is greatest at the *inflection point* with $m = [v_r] = [tl/(q_{\tilde{r}}\pi)]$ and falling as the distance along the arc $|d|$ increases, either towards the higher $m = [tl/((q_{\tilde{r}} - 1)\pi)]$ in the *distal* $P(\mathcal{R}_r)$ or towards the lower $m = [tl/((q_{\tilde{r}} + 1)\pi)]$ in the *proximal* $P(\mathcal{R}_r)$. Recalling $\ln(1 + x) = \sum_{n=1}^{\infty}(-1)^{n+1}n^{-1}x^n$ we have $\ln((m + l)/m) = \sum_{a=1}^{\infty}(-1)^{a+1}a^{-1}(l/m)^a$ allowing us to accept for *distal* structures, $t\ln((m + l)/m) \approx tl/m$ with $l|m$.

The change in the arguments between neighbouring tangential l^{th} vectors $(l|m)$ for any $P(\mathcal{R}_r)$ of any $\eta_l(s)$ is presented in the interval $(-\pi, \pi)$ as

$$\Delta\theta = q_{\tilde{r}}\pi - tl/m \text{ with } q_{\tilde{r}} \in \chi_l. \tag{55}$$

This gives the reduced angles $-\pi < \Delta\theta < 0$ in the *proximal* part of a $P(\mathcal{R}_r)$ and $0 < \Delta\theta < \pi$ in the *distal* part of a $P(\mathcal{R}_r)$. This relationship also applies to any two vectors $l \nmid m$ with $l - 1$ intervening vectors even though they are not tangential to $Y_r(u)$.

With m satisfying $l|m$ the radius of curvature near $\vec{m}$ is $\rho_{\mathcal{R}_r}(m) = lm^{-\sigma}/\Delta\theta$ giving $\rho_{\mathcal{R}_r}(m) = lm^{1-\sigma}/(mq_{\tilde{r}}\pi - tl)$. Briefly, we imagine using $lm^{-\sigma}$ and not $(l - 1)m^{-\sigma}$ to take up the interval vacated by the $\vec{m}$ where $l \nmid m$. The distance d from the *inflection point* along $Y_r(u)$ into the *distal* spiral is $d > 0$ and into the *proximal* spiral $d < 0$. We now seek $x_r(d) = -\sum \Delta d \cos(\theta)$ and $y_r(d) = -\sum \Delta d \sin(\theta)$ as shown in Figure 12, before letting Δd tend to zero using s as a dummy variable and $\lim_{\Delta d \to 0} \Delta d = ds$.

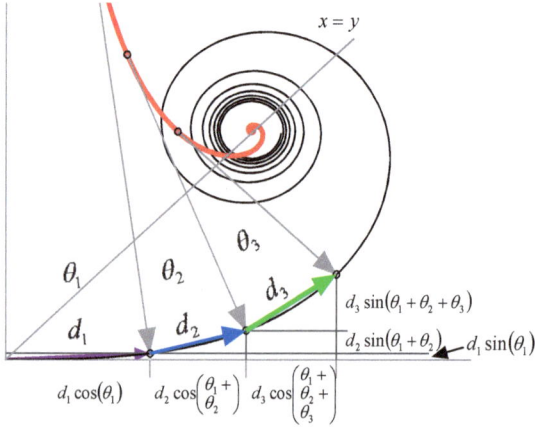

Figure 12. Shows $\Delta\theta$ as a series of θ_i such that $\theta = \sum \theta_i$, and Δd as a series of d_i such that $d = \sum d_i$. This shows $x_r(d) = -\sum d_i \cos(\theta)$ and $y_r(d) = -\sum d_i \sin(\theta)$; note that $d_i < 0$ in the *proximal* spiral.

Knowing that $d = \int \mu^{-\sigma} d\mu$ implies $d = \mu^{1-\sigma}/(1-\sigma)$ helps us summate the arc from $\vec{v}_r$ to $\vec{n}$ as $d = \sum_{v_r}^{n} m^{-\sigma} \approx (v_r^{1-\sigma} - n^{1-\sigma})/(1-\sigma)$, and then since $ds = \mu^{-\sigma} d\mu$ we can parametise in u rather than d with

$$x_r(d) = -\int_{s=0}^{s=d} \cos(\theta) ds \Rightarrow x_r(u) = -\int_{\mu=v_r}^{\mu=u} \mu^{-\sigma} \cos(\theta) d\mu$$

$$y_r(d) = -\int_{s=0}^{s=d} \sin(\theta) ds \Rightarrow y_r(u) = -\int_{\mu=v_r}^{\mu=u} \mu^{-\sigma} \sin(\theta) d\mu. \tag{56}$$

For $l = 2$ all $\vec{m}$ of a $P(\mathcal{R}_r)$ are tangential to $Y_r(u)$, but when $l \geq 2$ only those $\vec{m}$ where $l|m$ are tangential. However, we can re-imagine the geometry of a $P(\mathcal{R}_r)$ by pretending that $\Delta\theta(m)$, at any m, is $1/l \times$ the difference between the $\arg(\vec{m})$ and $\arg(\vec{m+l})$ and that $|\vec{m}|$ where $l|m$ is not of magnitude $(l-1)m^{-\sigma}$ but rather $m^{-\sigma}$ and that the vectors where $l \nmid m$ rather than being off to one side also lie along $Y_r(u)$. We then expect $\theta(u) = \frac{q_{\bar{r}} \pi}{l}(u - v_r) - t\ln\left(\frac{u}{v_r}\right)$ with its derivation illuminating the relationship between the paired spirals of a $P(\mathcal{R}_r)$ and paired Euler's spirals.

3.3. $\theta(u) = \frac{q_{\bar{r}} \pi}{l}(u - v_r) - t\ln\left(\frac{u}{v_r}\right)$

Let M_d (d for *distal*) be the largest m in the sequential set $\{m \in \mathbb{N}_1, m \in \mathcal{R}_r : m > v_r\}$ associated with part of a *distal* $P(\mathcal{R}_r)$ starting at the *inflection point*. Similarly, let M_p (p for *proximal*) be the smallest m in a sequential set $\{m \in \mathbb{N}, m \in \mathcal{R}_r : m < v_r\}$ associated with part of a *proximal* $P(\mathcal{R}_r)$ ending at the *inflection point*. The index $i \in \mathbb{Z}$ counts the position of an m from the *infection point*; $i = m - [v_r]$ with $i > 0$ in the *distal* and $i < 0$ in the *proximal* $P(\mathcal{R}_r)$.

Starting from the *inflection point* we seek $\theta(M)$ a summation of $1/l \times$ the difference between the arguments of $\vec{m}$ that are l apart. This relationship applies even if $l \nmid m$ allowing $\theta(M)$ for both M_d and M_p to be;

$$\theta(M) = \sum_{i=0}^{i=M-[v_r]} \left(\frac{q_{\bar{r}}\pi}{l} - \frac{t}{l}\ln\left(\frac{[v_r] + i + l}{[v_r] + i}\right)\right). \tag{57}$$

To move towards a smooth curve and an integral we let $\mu \in \mathbb{R}$ represent $m \in \mathbb{N}_1$ as in $m = [\mu]$ and define $\Delta\mu$ by letting $\mu = v_r + i\Delta\mu$. The variable u is our target and μ its related placeholder or dummy variable that vanishes on integration. Our summated $\theta(u) = \sum \Delta\theta$ is now explicitly

$$\theta(u) = \sum_{i=0}^{i=\left[\frac{u-v_r}{\Delta u}\right]} \left(\frac{\Delta\mu q_{\bar{r}}\pi}{l} - \frac{t}{l}\ln\left(\frac{v_r + i\Delta\mu + l\Delta\mu}{v_r + i\Delta\mu}\right)\right). \tag{58}$$

Replacing v_r with $\mu - i\Delta\mu$ gives

$$\theta(u) = \sum_{i=0}^{i=\left[\frac{u-v_r}{\Delta u}\right]} \left(\frac{\Delta\mu q_{\bar{r}}\pi}{l} - \frac{t}{l}\ln\left(1 + \frac{l\Delta\mu}{\mu}\right)\right). \tag{59}$$

Given that $\ln(1+x) = \sum_{n=1}^{\infty}(-1)^{n+1}n^{-1}x^n$ we consider

$\ln\left(1 + \frac{l\Delta\mu}{\mu}\right) = \frac{l\Delta\mu}{\mu} + \mathcal{O}(\mu^{-2})$, and so, accepting

$$\theta(u) = \sum_{i=0}^{i=\left[\frac{u-v_r}{\Delta\mu}\right]} \left(\frac{\Delta\mu q_{\bar{r}}\pi}{l} - \frac{t\Delta\mu}{\mu}\right) \tag{60}$$

we proceed to infinitesimals with

$$\theta(u) = \int_{v_r}^{u} \frac{q_{\bar{r}}\pi}{l} d\mu - \int_{v_r}^{u} \frac{t}{\mu} d\mu. \tag{61}$$

Integrating Equation 61 gives the result that we anticipated when comparing tangents at u and v_r, namely $\theta(u) = \frac{q_{\bar{r}}\pi}{l}(u - v_r) - t\ln\left(\frac{u}{v_r}\right)$. Then, since $v_r = tl/q_{\bar{r}}\pi$ for large t we have

$$\theta(u) = t\left(\frac{u}{v_r} - 1\right) - t\ln\left(\frac{u}{v_r}\right). \tag{62}$$

We now apply the power series $\ln(1+x) = \sum_{n=1}^{\infty}(-1)^{n+1}n^{-1}x^n$ and see that

$$\ln\left(\frac{u}{v_r}\right) = \ln\left(1 + \frac{u-v_r}{v_r}\right)$$

$$= \left(\frac{u}{v_r} - 1\right) - \frac{1}{2}\left(\frac{u}{v_r} - 1\right)^2 + \frac{1}{3}\left(\frac{u}{v_r} - 1\right)^3 - \frac{1}{4}\left(\frac{u}{v_r} - 1\right)^4 + \cdots \tag{63}$$

and ask ourselves if we can accept the two emboldened terms as sufficient, so producing the helpful

$$\theta(u) = \frac{t}{2}\left(\frac{u}{v_r} - 1\right)^2 = \frac{t}{2v_r^2}(u - v_r)^2. \tag{64}$$

This seems helpful since it takes us towards a robust analogy with paired Euler's spirals whose behaviour we understand—but are the first two terms of Equation 63 sufficient? Can we discard terms $n^{-1}(u/v_r - 1)^n$ for $n > 2$? The following analysis says that we can.

If $\theta(u)$ follows Equation 62 then $Y_r(u)$ models a $P(\mathcal{R}_r)$ throughout its sequential vectors and allows estimation of the orientation and magnitude of the *principal-axis* through location of its foci using an approximation method similar to that illustrated in Figure 9(b) on page 30.

If $\theta(u)$ follows Equation 64 then $Y_r(u)$ models a $P(\mathcal{R}_r)$ for a sub-set of sequential vectors $\{\vec{m}\} \subset \mathcal{R}_r$ (Equation 19) for which m/v_r is close to unity, lying either side of $\vec{v}_r$. This region connects the paired spirals and locates their first few turns. Outside of that set the curve and the pathway part company. Fortunately, this is only a problem at low values of t as illustrated in Figure 13.

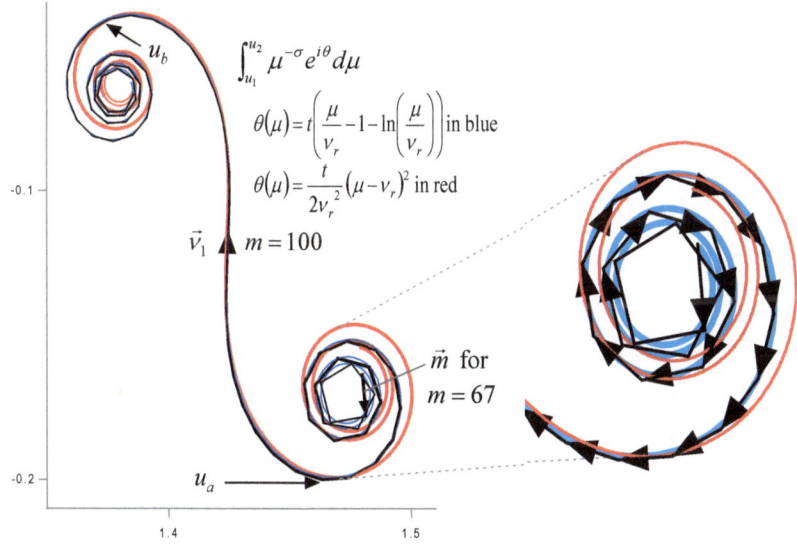

Figure 13. In black the vectors of $P(\mathcal{R}_1)$ for $\eta_2(s)$ with $t = 100\pi$ for $\sigma = 1$ from the 67th to the 100th vector in the *proximal* part and the 100th to the 146th vector in the *distal* part. The curves follow $\int_{66}^{u} \mu^{-\sigma} e^{i\theta} d\mu$ with $\theta(\mu) = t\left(\frac{\mu}{v_r} - 1\right) - t \ln\left(\frac{\mu}{v_r}\right)$ in blue and $\theta(\mu) = \frac{t}{2v_r^2}(\mu - v_r)^2$ in red, with u allowed to rise from 66 to 147. Inset: an enlarged view of the *proximal* pseudo-spiral.

When t is low the ratio m/v_r throughout a $P(\mathcal{R}_r)$ may be sufficiently removed from unity that we cannot discard terms $n^{-1}(u/v_r - 1)^n$ for $n > 2$. In Figure 13 an estimate of the location of the *distal* focal point would be poor using $\theta(\mu) = \frac{t}{2v_r^2}(\mu - v_r)^2$ with the ratio u/v_r being sufficiently close to unity only as far as u_b/v_r and u_a/v_r, which only takes us towards the first turns of the spirals, and not sufficiently far into them to locate the foci.

Moving to a higher, but still modest $t = 100,000\pi$, Figure 14 below, plots $P(\mathcal{R}_1)$ and $P(\mathcal{R}_{27})$ to show that both structures allow us to discard terms $n^{-1}(u/v_r - 1)^n$ for $n > 2$. In Figure 14(b) the better fit of $\int \mu^{-\sigma} e^{i\theta} d\mu$ using $\theta(\mu) = \frac{t}{2v_r^2}(\mu - v_r)^2$ in red, in comparison with that in Figure 13 (on page 35), is easily understood since m/v_r throughout the $P(\mathcal{R}_r)$ remains close to unity and this is maintained as r rises. Remember, we only need r to rise until m falls to κl. In Figure 14(b) with $r = 27$ we have m/v_r ranging from 1.009 to 0.99. This example illustrates why confidence in accepting the first two terms of Equation 63 rises as t rises.

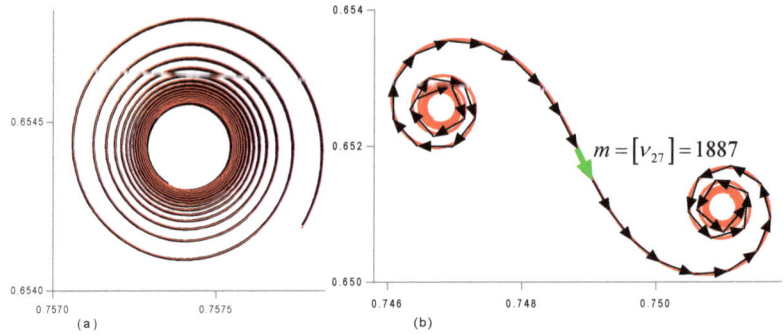

Figure 14. (a) In black, $P(\mathcal{R}_1)$ for $\eta_2(s)$ with $t = 100,000\pi$, $\sigma = 1$, from the 97,476th to the 99,243rd vector in the *proximal* spiral, and in red $\int \mu^{-\sigma} e^{i\theta} d\mu$ using $\theta(\mu) = \frac{t}{2v_r^2}(\mu - v_r)^2$. (b) $P(\mathcal{R}_{27})$ i.e. $q_{\bar{r}} = 106$ from $m = 1869 \to q = 107$, to $m = 1904 \to q = 105$. The integral and spirals follow γ from -3.886 to 3.886.

Theta, based on two terms as $\theta(\mu) = \frac{t}{2v_r^2}(\mu - v_r)^2$ now informs $Y_r(u)$ for *distal* $u > v_r$ and *proximal* $u < v_r$ following $P(\mathcal{R}_r)$ with $x_r(u) = -\int_{v_r}^{u} \mu^{-\sigma} \cos(\theta) d\mu$ and $y_r(u) = -\int_{v_r}^{u} \mu^{-\sigma} \sin(\theta) d\mu$, which run over the range (v_r, u) in Equations 65 and 66;

$$x_r(u) = -\int_{v_r}^{u} \mu^{-\sigma} \cos\left(\frac{t}{2v_r^2}(\mu - v_r)^2\right) d\mu \qquad (65)$$

$$\text{and } y_r(u) = -\int_{v_r}^{u} \mu^{-\sigma} \sin\left(\frac{t}{2v_r^2}(\mu - v_r)^2\right) d\mu. \qquad (66)$$

In Euler's spiral $\theta(d) \propto d^2$ whereas $\theta(u) \propto (u - v_r)^2$, and though $(u - v_r)$ grows with u we have to consider $\mu^{-\sigma}$. Equations 65 and 66 are similar to Euler's spirals when $\sigma = 0$ since $\partial d = \partial u$, $d = u - v_r$ and $\mu^{-\sigma} = 1$. When $u > v_r$ we can let $u \to \infty$ but when $u < v_r$ the integral is poorly behaved when $u \to 1$. Fortunately, u/v_r need not stray far from unity.

We now compare *proximal* and *distal* spirals in $Y_r(u)$ when $\sigma > 0$ and make some substitutions. We have $v_r = tl/(q_{\bar{r}}\pi)$ and when $l > \kappa$ we have $q_{\bar{r}}/2 = r$ giving $t = 2r\pi v_r/l$. Into Equations 65 and 66 it is now natural to let $c = \sqrt{t/2}(\mu/v_r - 1)$ and consequently $dc = \sqrt{lv_r/(r\pi)}\, du$. This substitution changes the limits, with $c = 0$ at the *inflection point*, and brings important terms outside of the integration. We now have c as the dummy variable and will temporarily use $\pm \gamma$ for symmetrical limits of integration in $Y_r(\gamma)$ and use for convenience $\mathbb{C}$ instead of $\mathbb{R}^2$;

$$Y_r(\gamma) = -\sqrt{l/(r\pi)}\, v_r^{(1/2-\sigma)} \int_{-\gamma}^{\gamma} \left(c\sqrt{2/t} + 1\right)^{-\sigma} e^{ic^2}\, dc. \tag{67}$$

The *principal-axis* of these paired spirals is the vector $\vec{Y}_r$ with components;

$$\mathrm{Re}(\vec{Y}_r) \approx -\sqrt{l/(r\pi)}\, v_r^{(1/2-\sigma)} \int_{-\gamma_1}^{\gamma_1} \left(c\sqrt{2/t} + 1\right)^{-\sigma} \cos(c^2)\, dc$$

$$\mathrm{Im}(\vec{Y}_r) \approx -\sqrt{l/(r\pi)}\, v_r^{(1/2-\sigma)} \int_{-\gamma_2}^{\gamma_2} \left(c\sqrt{2/t} + 1\right)^{-\sigma} \sin(c^2)\, dc$$

with limits $\gamma_1 = \sqrt{\dfrac{n\pi}{2}}, \gamma_2 = \sqrt{\dfrac{(n+1)\pi}{2}}, n \in \mathbb{E}$ and $4 < n \ll t/\pi$. \hfill (68)

Once n exceeds 4 any further rise makes only minimal changes in $|\vec{Y}_r|$ and $\arg(\vec{Y}_r)$ (see Figure 9(b) page 30) as long as $\sqrt{(n+1)\pi/t} \ll 1$ which is easily achieved, even at modest values of t.

Importantly, Equation 67 frees the integrands from r and v_r and gives $|\vec{\mathcal{R}}_r|$ exactly as in Equation 22 (page 19). When $\sigma = 0$ the spirals of a $P(\mathcal{R}_r)$ are congruent with the paired spirals of Euler over matching intervals and consequently the magnitude of $|\vec{\mathcal{R}}_r| = \sqrt{lv_r/r}$ when $\sigma = 0$.

When $\sigma \neq 0$ the term $\left(c\sqrt{2/t} + 1\right)^{-\sigma} \neq 1$, but for values of $c \ll t$ the $\lim_{t \to \infty} \left(\sqrt{2/t}\, c + 1\right)^{-\sigma} \to 1$ and is close to 1 at high values of t and so the spirals encoded in the integrals resemble the Fresnel integrals $\int_{-\infty}^{\infty} \cos(c^2)\, dc = \int_{-\infty}^{\infty} \sin(c^2)\, dc = \sqrt{\pi/2}$. However, we can do better than this by defining two transcendental integrals as functions of σ and t with limits as before,

$$C(\sigma,t) = -\int_{-\gamma_1}^{\gamma_1} \left(c\sqrt{2/t}+1\right)^{-\sigma} \cos(c^2)\, dc \tag{69}$$

$$S(\sigma,t) = -\int_{-\gamma_2}^{\gamma_2} \left(c\sqrt{2/t}+1\right)^{-\sigma} \sin(c^2)\, dc. \tag{70}$$

These then allow generic paired spirals to be imagined with a *principal-axis* $\vec{Y}_g(\sigma,t)$,

$$\vec{Y}_g(\sigma,t) = C(\sigma,t) + iS(\sigma,t), \tag{71}$$

whose magnitude and argument are simply $|\vec{Y}_g| = \sqrt{C^2+S^2}$ and $\arg(\vec{Y}_g) = \operatorname{atan}\left(\frac{S}{C}\right)$ and allowing

$$\vec{Y}_r(\sigma,t) = \sqrt{\frac{l}{r}} v_r^{(1/2-\sigma)} \vec{Y}_g(\sigma,t), \tag{72}$$

with $\frac{\partial}{\partial \sigma}\arg(\vec{Y}_r) = \frac{\partial}{\partial \sigma}\arg\left(\vec{Y}_g(\sigma,t)\right)$. It is now necessary to prove that

$$\frac{\partial}{\partial \sigma}\arg\left(\vec{Y}_g(\sigma,t)\right) = 0.$$

3.4. A note on the Fresnel integrals

A formal proof of the Fresnel integrals takes a contour integral of e^{-t^2} around a sector that bisects Quadrant I. A less rigorous appreciation starts with half the Gaussian integral $\int_0^\infty e^{-t^2} dt = \sqrt{\pi}/2$; consider $I = \int_0^\infty e^{it^2} dt$ and let $t = xe^{i\pi/4}$ so that $dt = e^{i\pi/4} dx$ and since $\left(e^{i\pi/4}\right)^2 = i$ we have $it^2 = -x^2$ giving $I = e^{i\pi/4}\int_0^\infty e^{-x^2} dx = e^{i\pi/4}\frac{\sqrt{\pi}}{2}$. It then follows that $I = \int_0^\infty \cos(t^2)\, dt + i\int_0^\infty \sin(t^2)\, dt = \left(1/\sqrt{2}+i/\sqrt{2}\right)\sqrt{\pi}/2$ giving the *principal-axis* of paired Euler spirals a length of $\sqrt{\pi}$ at an argument of $\pi/4$. However, Equations 69 and 70 are not as straightforward to solve.

3.5. The principal-axis and argument of $\vec{Y}_g(\sigma,t)$

If σ rises or falls from 0 *proximo-distal shrinkage* in a P($\mathcal{R}_r$), now encoded in Equations 69 and 70, distorts the *proximal* and *distal* spirals and they lose their congruence. In the range $-1 \leq \sigma \leq 1$ the incongruence is subtle and so we go outside the *critical strip* to illustrate it.

For simplicity, Figure 15 plots $x(\gamma) = \int_{-\gamma}^{\gamma} \left(c\sqrt{2/t} + 1\right)^{-\sigma} \cos(c^2) \, dc$ and $y(\gamma) = \int_{-\gamma}^{\gamma} \left(c\sqrt{2/t} + 1\right)^{-\sigma} \sin(c^2) \, dc$ placing the *proximal* focus in the bottom left. Four values of σ demonstrate the stability of the *principal-axis*.

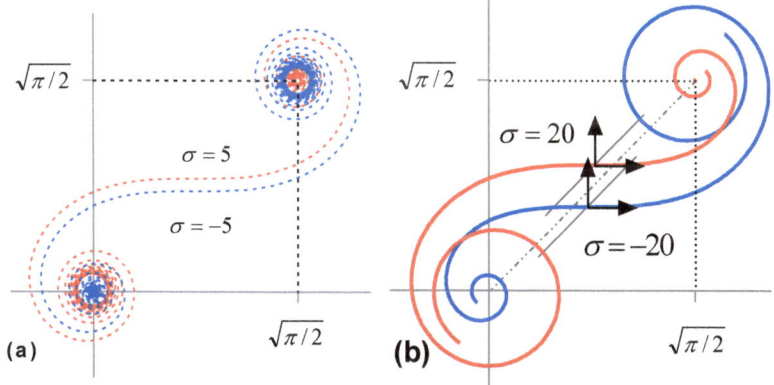

Figure 15. In red $\sigma > 0$ and in blue $\sigma < 0$ with **(a)** $\sigma = \pm 5$ and **(b)** $\sigma = \pm 20$ for $v_r = 4000 \Rightarrow t = 12566.3$. The *inflection point* at $\gamma = 0$ lies to the left of the *principal-axis* when $\sigma > 0$ and to the right when $\sigma < 0$. The *proximal* focus is located in the bottom left.

The axis is stable because changes to the *proximal* and *distal* spirals are beautifully balanced, as proved below.

3.6. A double sized single Euler spiral

We combine the *proximal* spiral and a negated *distal* spiral of paired Euler spirals into one double-sized single Euler spiral designated $2E(d)$ with $d = \infty$ giving $2\int_0^\infty e^{is^2} ds = \sqrt{\pi/2}\,(1 + i)$. The doubling makes $2\vec{E}$ equivalent to the *principal-axis* of the paired Euler spirals, which may seem pointless. However, it permits a direct comparison with a single spiral constructed from the *proximal* and *distal* spirals that sit behind $\vec{Y}_g(\sigma)$ with $\sigma \neq 0$.

The parametised double-sized Euler spiral in Figure 16, has $y(d) = \text{Im}\left(2\int_0^d e^{is^2} ds\right)$ and $x(d) = \text{Re}\left(2\int_0^d e^{is^2} ds\right)$ plotted against d showing both functions crossing the dotted horizontal line $\sqrt{\pi/2}$ close to points of inflection with a phase difference which is rapidly asymptotic to $\pi/2$. In $x(d)$ the points of inflection are at $d = \sqrt{n\pi/2}$ when $n \in \mathbb{E}_{>2}$, since $\frac{d^2}{dd^2} \int_0^d \cos(s^2) ds = -2d \sin(d^2)$. In $y(d)$, the points of inflection are at

$d = \sqrt{(n+1)\pi/2}$ with $n \in \mathbb{E}_{>2}$, since $\frac{d^2}{dd^2}\int_0^d \sin(s^2)ds = 2d\cos(d^2)$. These permit the approximations $2\int_0^\infty \cos(s^2)\,ds \approx 2\int_0^d \sin(s^2)\,ds = \sqrt{\pi/2}$ when $d = \sqrt{n\pi/2}$ with $n \in \mathbb{E}$ and $2\int_0^\infty \sin(s^2)\,ds \approx 2\int_0^d \sin(s^2)\,ds = \sqrt{\pi/2}$ when $d = \sqrt{(n+1)\pi/2}$ with $n \in \mathbb{E}$ and $2 < n \ll \infty$.

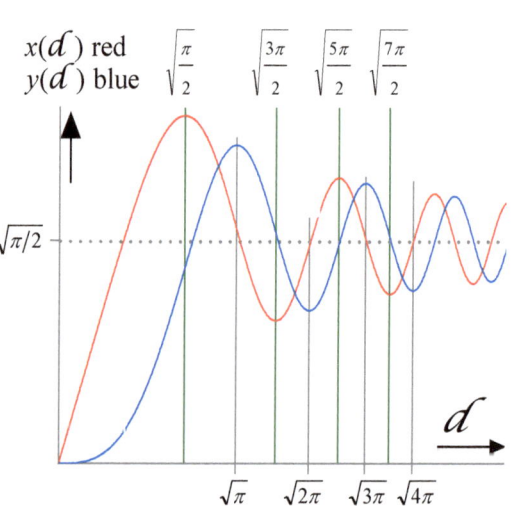

Figure 16.
The parameters of $2E(d)$ being
$x(d) = 2\int_0^d \cos(s^2)\,ds$ in red
and
$y(d) = 2\int_0^d \sin(s^2)\,ds$ in blue showing
$x(\infty) \cong \int_0^{\sqrt{n\pi/2}} \cos(s^2)\,ds$ for $n \in \mathbb{E}$ and
$y(\infty) \cong \int_0^{\sqrt{n\pi/2}} \sin(s^2)\,ds$ for $n \in \mathbb{O}$.

We now seek $\vec{H}(\sigma)$ which approximates to a $\vec{Y}_g$. Our $\vec{H}(\sigma)$ is a vector from the origin to the "centre of convergence" of a solitary spiral being the sum of the *proximal* spiral and a negated *distal* spiral of $Y_g(\gamma)$ but whose real and imaginary parts conclude at different limits of integration. The real part runs to $\gamma = \sqrt{n\pi}$ with $n \in \mathbb{E}$ and $2 < n \ll t/2$ and the imaginary part finishes after a further $\pi/2$. The vector $\vec{H}(\sigma) \in \mathbb{C}$ is defined as

$$\vec{H}(\sigma) = \int_0^\gamma \left(\left(c\sqrt{2/t}+1\right)^{-\sigma} + \left(1-c\sqrt{2/t}\right)^{-\sigma}\right)\cos(c^2)\,dc$$

$$+ i\int_0^{\gamma+\frac{\pi}{2}} \left(\left(c\sqrt{2/t}+1\right)^{-\sigma} + \left(1-c\sqrt{2/t}\right)^{-\sigma}\right)\sin(c^2)\,dc$$

$$\gamma = \sqrt{n\pi/2} \text{ with } n \in \mathbb{E} \text{ and } 2 < n \ll \sqrt{t/2}. \tag{73}$$

At the limit of the *critical strip* when $\sigma = 1$, where the discrepancy from $2E(d)$ causes the greatest anxiety, we have

$$\lim_{\sigma \to 1}\left(\left(c\sqrt{2/t}+1\right)^{-\sigma} + \left(1-c\sqrt{2/t}\right)^{-\sigma}\right) = 2 + 4c^2/(t-2c^2). \tag{74}$$

Consequently, if $c \ll \sqrt{t/2}$ the sum $\left(c\sqrt{2/t}+1\right)^{-\sigma} + \left(1-c\sqrt{2/t}\right)^{-\sigma} \approx 2$ and so $\vec{H}(\sigma) \approx 2\int_0^\infty e^{ic^2}dc = \sqrt{\pi/2}\,(1+i)$ giving $|\vec{H}(\sigma)| = \sqrt{\pi}$ for all σ and $\frac{d}{d\sigma}\arg\left(\vec{H}(\sigma)\right) = 0$, see Figure 17.

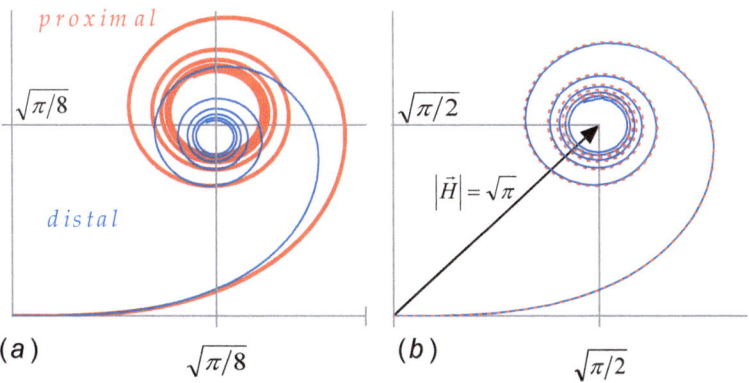

Figure 17. In **(a)** $\int_0^\gamma \left(c\sqrt{2/t}+1\right)^{-\sigma} e^{ic^2}dc$ in blue representing the distal P($\mathcal{R}_r$), and in red $\int_0^\gamma \left(1-c\sqrt{2/t}\right)^{-\sigma} e^{ic^2}dc$ representing the proximal P($\mathcal{R}_1$) for $\sigma = 20$, $t = 1{,}256{,}637.061$. In **(b)** on a different scale, $\int_0^\gamma \left(\left(c\sqrt{2/t}+1\right)^{-\sigma} + \left(1-c\sqrt{2/t}\right)^{-\sigma}\right) e^{ic^2}dc$ in dashed red and in blue a single Euler's spiral $2E(d) = 2\int_0^d e^{is^2}ds$. Both panels show d and γ over the interval 0 to 6.

This section has proved that over the *critical strip*, for high values of t,

$$\text{Lemma (3)} \quad \frac{\partial \arg(\vec{\mathcal{R}}_r)}{\partial \sigma} = 0$$

and

$$\text{Lemma (4)} \quad |\vec{\mathcal{R}}_r| = \sqrt{l/r}\, v_r^{1/2-\sigma}.$$

4. The Breaking of Symmetry

4.1. The function $\zeta_\tau(\rho) = 0$ and $\lim_{\sigma \to \infty}(\zeta_\tau(s)) = 1 + 0i$

In $\eta_{l,n}(\rho) \approx 0$ the incorporated series $\zeta_\tau(\rho) = 0$. If $\zeta_\tau(\rho)$ were to play a role in generating a ρ_u when $\sigma \neq 1/2$, then it must do so at all values of l and yet $\zeta_\tau(s)$ is insensitive to l, whilst $\ell_{l,n}(s)$ and $h_{l,r}(s)$ are affected by l. For these reasons, if there were a $\eta_l(\rho_u) = 0$ there has to be a $\zeta_\tau(\rho_u) = 0$.

4.2. In Euler's zeta, the final pseudo-spiral ends where divergence starts at the tau vector

Where do we place the focal point of the final pseudo-spiral of $\zeta_n(s)$? For high t we can use $Y_1(\gamma)$ but is there another way? Such a point could be $\phi(s)$ employing the *tau* vector. Our $\phi(s)$ could generate a point in $P(\zeta_n(s))$ half-way along $\vec{\tau}$, or one based on an average of $\vec{\tau}$ with a neighbour but a refined mechanism could be to use the diverging spiral when $m > \tau$, see Figure 18.

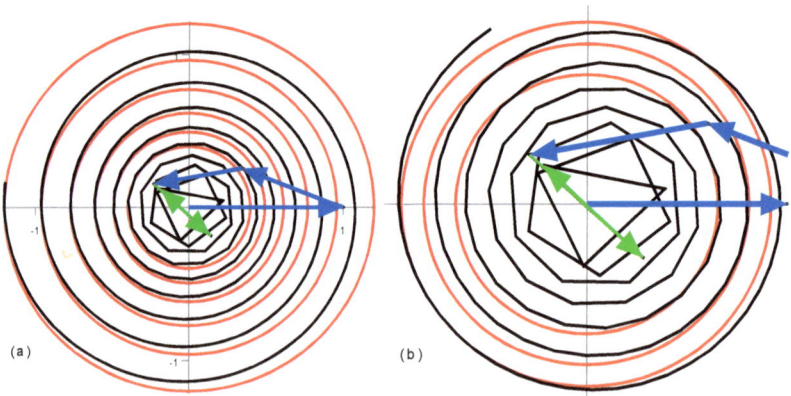

Figure 18. (a) In blue, the first 3 vectors of $\zeta_\tau(1/2 + 14.1347i)$ are shown, and in green, the 4th and the *tau* vectors pass close to zero. The growing black spiral is the start of divergence and the red circles are all centred at 0. The spiral of divergence crosses each red circle only once. **(b)** Illustration to show how $P(\zeta_n(s))$ in black would cross or touch a circle centred at zero twice when $s \notin \{\rho\}$ in spite of $\vec{\tau}$ (green) now passing through zero. This example has $\sigma = 0.55$ and $t = 14.1347$, a similar plot would apply if $\sigma = 1/2$ and $t = 14.23$.

Dirichlet's *eta*, obviates the need for a $\phi(s)$, but the points above are made to emphasis that (1) there is no reason to believe that a $P(\zeta_n(\rho_{u_\alpha}))$

would have a vector $\vec{\tau}$ that did not cross close to zero, nor (2) a divergent spiral whose $\phi(\rho_{u_\alpha})$ was not centred at zero. In these respects, $P(\zeta_n(\rho_{u_\alpha}))$ would share the features shown in Figure 18(a) that follow for all elements of $\{\rho_i\}$. For these reasons, it is axiomatic that for high values of t, where RH might fail, we can say that, if $\eta_l(\rho_{u_\alpha}) = 0$ then for $\zeta_\tau(\rho_{u_\alpha})$ we have a $\phi(\rho_{u_\alpha}) = 0$ and we lose no force of argument by considering that if *tau* is the subscript we can understand that $\zeta_\tau(\rho_{u_\alpha}) = 0$, or at least give meaning to it acknowledging an appropriate averaging function. The arguments used here apply if $Y_r(\gamma)$ were used instead.

4.3. Symmetry in Euler's zeta

Figure 19(a) shows $P(\zeta_n(t_{901}))$ in black, with arrows for the first five $\vec{m}$ of the *proximal* pathway. The symmetries are striking since $\psi_0 \sim \pi/2$. There are 14 *proximal* vectors emphasized in blue, and many *distal* vectors forming paired pseudo-spirals $P(\mathcal{R}_r)$ ending back at the origin; $P(\mathcal{R}_3)$ and $P(\mathcal{R}_1)$ are labeled. The *distal* pathway has 14 red $\vec{\mathcal{R}}_r$ laid upon it. These mirror the blue vectors and the first five of $P(h_{1,\kappa}(s))$ have red arrow heads. The black $P(\zeta_n(s))$ diverges as implied by the grey clockwise spiral representing $m > \tau$. Figure 19(a) illustrates the relationship

$$P(\zeta_\tau(\rho_i)) = P(\zeta_\kappa(\rho_i)) - P(h_{1,\kappa}(\rho_i)) \tag{75}$$

or to be precise and to accommodate the intersection of the *kappa* vectors

$$P(\zeta_\tau(\rho_i)) = P(\zeta_{\kappa-1}(\rho_i)) + x\vec{\kappa} - x\vec{\mathcal{R}}_\kappa - P(h_{1,\kappa-1}(\rho_i)). \tag{76}$$

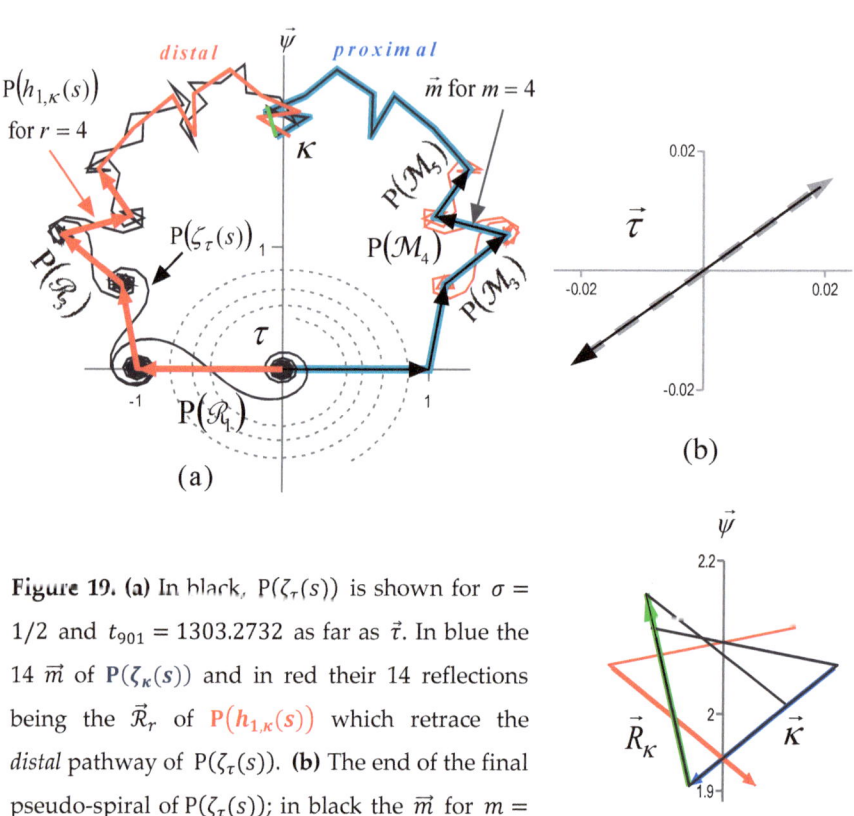

Figure 19. (a) In black, $P(\zeta_\tau(s))$ is shown for $\sigma = 1/2$ and $t_{901} = 1303.2732$ as far as $\vec{\tau}$. In blue the 14 $\vec{m}$ of $P(\zeta_\kappa(s))$ and in red their 14 reflections being the $\vec{\mathcal{R}}_r$ of $P(h_{1,\kappa}(s))$ which retrace the *distal* pathway of $P(\zeta_\tau(s))$. **(b)** The end of the final pseudo-spiral of $P(\zeta_\tau(s))$; in black the $\vec{m}$ for $m = 414$ and in dashed grey $\vec{\tau}$ for $m = 415$. **(c)** The $\vec{m}$ for which $m = \kappa$ is shown in blue as $\vec{\kappa}$ and the $\vec{m}$ for which $m = \kappa + 1$ is shown in green. In red $\vec{\mathcal{R}}_\kappa$ the κth vector of $P(h_{1,\kappa}(s))$ intersects with $\vec{\psi}$ where the paired vectors cross one another.

In Figure 19(a) two $P(\mathcal{M}_m)$, made of many $\vec{\mathcal{R}}_r$, are in red. Divergence in $P(\zeta_n(s))$, shown as a dashed grey spiral, starts as soon as $\Delta\theta$ falls below π after the *tau* vector $\vec{\tau}$, see Figure 19(b). In Figure 19(c), the junction of *proximal* and *distal* parts of the pathway are redrawn with the $\vec{\kappa}$ in blue. The *inflection point* in $P(\mathcal{R}_1)$ is near $m = \left[(e^{2\pi/t} - 1)^{-1}\right] = \kappa^2$ as shown in Figure 20 (opposite).

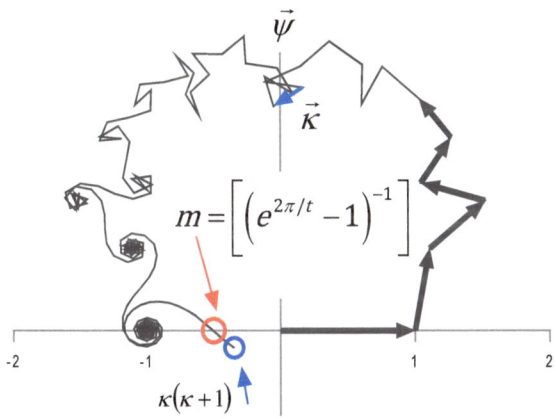

Figure 20. P($\zeta_n(s)$) for the 901st nontrivial zero. The red circle and arrow are at the *inflection point*.

Later we see that when $l \geq \kappa + 1$ there is symmetry in P$(\eta_{l,n}(s))$ and if $l = \kappa + 1$ then $\zeta_{\kappa l}(\rho)$ is near the *inflection point*. If $l = [\tau/\kappa]$ we have $\zeta_{\kappa l}(\rho_i) = \zeta_\tau(\rho_i) \approx 0$ and if $\{\rho_u\} \neq \emptyset$ we expect $\zeta_\tau(\rho_u) \approx 0$ with $\sigma \neq 1/2$.

4.4. Unique intersections with $\vec{\psi}$ in Euler's zeta: the function $\lambda_1(s,x)$

The final vectors of P($\zeta_\kappa(s)$) and P($h_{1,\kappa}(s)$) are $\vec{\kappa}$ and $\vec{\mathcal{R}}_\kappa$, and they intersect at $\vec{\psi}$ when Im(s) is close to Im(ρ_i), but all intersections have to be with $\sigma = 1/2$ to be in symmetry. Without a symmetrical intersection of the *kappa* vectors there can be no *zero*. The position along the *kappa* vector at which intersection takes place is $x \in \mathbb{R}$ with $0 < x \leq 1$. The value of x is the same for $\vec{\kappa}$ and $\vec{\mathcal{R}}_\kappa$. For *Euler's zeta* $l = 1$ and we define $\lambda_1(s,x) \in \mathbb{C}$ as the difference between the two pathways. Formally, *lambda* is a vector $\vec{\lambda}_1(s,x)$,

$$\vec{\lambda}_1(s,x) = \vec{\zeta}_{\kappa-1}(s) + x\vec{\kappa} - x\vec{\mathcal{R}}_\kappa - \vec{h}_{1,\kappa-1}(s), \qquad (77)$$

but no confusion is created by the notation $\lambda_1(s,x) \in \mathbb{C}$ giving an approximation to Euler's *zeta* of

$$\zeta_\tau(s) \approx \lambda_1(s,x). \qquad (78)$$

If we think of the pathways, we have $\vec{\lambda}_1(s,x)$ being the difference between the ends of two pathways in $\mathbb{R}^2$

$$\vec{\lambda}_1(s,x) = P(\zeta_{\kappa-1}(s) + x\vec{\kappa}) - P(h_{1,\kappa-1}(s) + x\vec{\mathcal{R}}_\kappa).$$

Empirically, $x = f(\acute{\kappa})$ is an oscillation whose frequency decreases as t rises, with x mostly satisfying $2/10 < x < 8/10$ (Appendix A5). Without an understanding of $x = f(t)$ Figure 21 uses $x = 1/2$ to illustrate how well $|\lambda_1(s,x)|$ locates the *zeros* of $\zeta(s)$ over an interval in t.

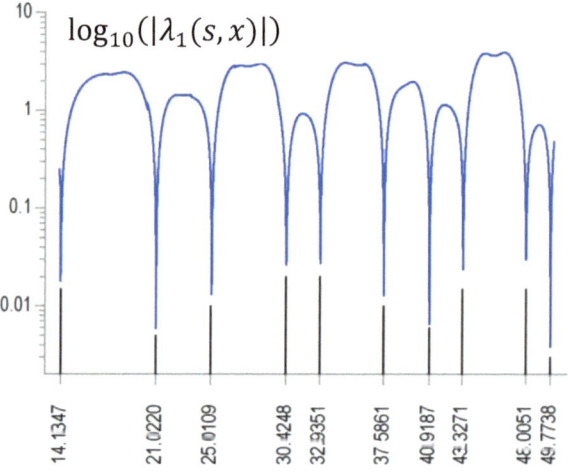

Figure 21. A plot of the log magnitude of $\lambda_1(s,x)$ for $\sigma = 1/2$ (using $x = 1/2$) over an interval embracing the first 10 *zeros* which are indicated with fine vertical lines.

Figure 21 shows minima in the magnitude of $\lambda_1(s,x)$ coinciding with the first 10 *zeros* of $\zeta(s)$. The function $\lambda_1(s,x)$ behaves well in this interval and it is of interest to identify the upper and lower bounds that could host zeros if $0 < x \leq 1$, rather than being fixed at $x = 1/2$.

4.5. The short finite series $\zeta_\kappa(s)$ and $h_{1,\kappa}(s)$ specify intervals in t which host nontrivial zeros

The short finite series $\zeta_\kappa(s)$ and $h_{1,\kappa}(s)$ can be used to determine the upper and lower bounds of an interval in t, which could host a *zero*. The interval in t is that for which there is overlap of the *kappa* vectors, either with each other, or with $\vec{\psi}$. These are formally the same but the latter is computationally simpler. An example is shown in Table 1.

Table 1. $\lambda_1(s,x) = 0$ for x over the interval 0 to 1

x	Lower bound	t_i	Upper bound	Δt Interval
[0→1]	1132489.1590	1132489.163	1132489.1711	0.0120
[1→0]	1132489.5795	1132489.585	1132489.5877	0.0081
[0→1]	1132490.1633	1132490.165	1132490.1697	0.0063

Intervals in t over which $\vec{\kappa}$ intersects with $\vec{\mathcal{R}}_\kappa$ or $\vec{\psi}$. [0→1] indicates that the lower bound relates to $x = 0$ and the upper bound to $x = 1$. [1→0] indicates that the lower bound relates to $x = 1$ and the upper bound to $x = 0$.

The intervals in t in Table 1 are conservative since the values $x = 0$ and $x = 1$ lie just inside the tabulated Lower and Upper bounds. This example shows that a short finite series, to a *kappa* of only 425, can identify narrow intervals in t capable of hosting *zeros* much higher up the imaginary axis.

4.6. Kappa and the number of nontrivial zeros

Each successive *kappa* makes a larger imaginary region accessible to *lambda* such that each value hosts more *zeros*. The left-hand axis of Figure 22 shows the total number of *zeros* (blue curve) and the right-hand axis (solid grey histogram) shows the number of *zeros* associated with each *kappa*.

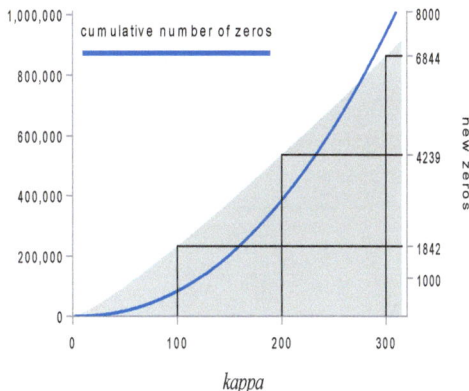

Figure 22.
The cumulative number of *zeros* is plotted in blue (left-hand axis) against *kappa* on the x axis.

The dense histogram in grey, (right-hand axis), shows the number of *zeros* associated with each *kappa*. For example, a *kappa* of 200 (specifying an interval in t of 2,512.68) covers 4,239 *zeros*. The interval starts at $t = 250,073.4$.

A short finite series can locate many *zeros*; a *kappa* of 423 is associated with an interval in t embracing 10,246 *zeros* from $t_i = 1,124,244.12$ to $t_i = 1,129,565.69$: these are the $1,985,173^{rd}$ zero and the $1,995,418^{th}$ *zero*, and each of these can be located to a small interval in t by $\lambda_l(s,x)$.

4.7. The behaviour of Euler's zeta when $\sigma \neq 1/2$

We return to Euler's *zeta*. Figure 23(a) (overleaf on page 48) illustrates the breaking of the symmetry that exists between the two finite series which track the behaviour of Euler's *zeta*. The pathways $P(\zeta_\kappa(s))$ and $P(h_{1,\kappa}(s))$ appear for the 901st *zero*, as in Figure 19 on page 44, but are also plotted for two values of $\sigma \neq 1/2$. Euler's *zeta* approximates as follows; $\zeta_\tau(s) \approx \zeta_\kappa(s) - h_{1,\kappa}(s)$. Figure 23(a) uses $\lambda_1(s,x)$ with $x = 0.77$ for plotting the larger vector equating to $\zeta_\tau(1/4 + it)$, hence this vector is not at the tip of the arrows representing $\vec{\kappa}$ and $\vec{\mathcal{R}}_\kappa$.

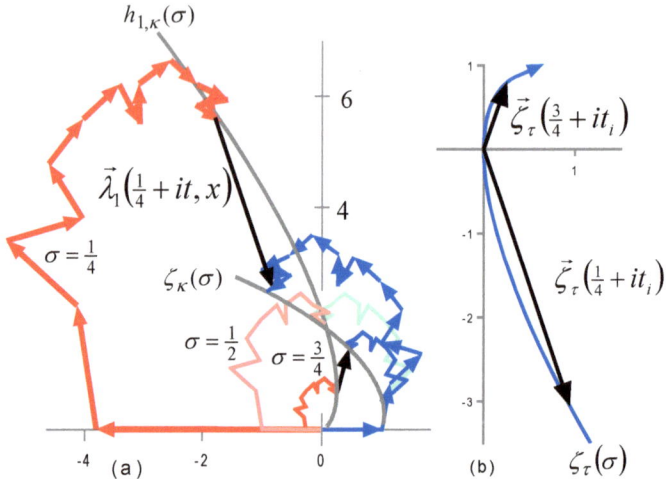

Figure 23. (a) $\lambda_1(s,x)$ with $P(h_{1,\kappa}(s))$ in red, $P(\zeta_\kappa(s))$ in blue, $\kappa = 14$, $t_i = t_{901}$. The light pathways with $\sigma = 1/2$ have intersecting $\vec{\kappa}$ vectors. Pathways for $\sigma = 1/4$ connect with a downward black vector $\vec{\lambda}_1(s,x)$. Pathways for $\sigma = 3/4$ connect with an upward black vector $\vec{\lambda}_1(s,x)$. The two grey curves represent $h_{1,\kappa}(\sigma)$ for fixed t ending at $(0, i0)$, and $\zeta_\kappa(\sigma)$ ending at $(1, i0)$. These curves cross at the *zero*. (b) $\zeta_\tau(\sigma)$ for fixed t in blue with σ rising in the direction of the blue arrow. Vectors $\vec{\zeta}_\tau(s)$ equate with the comparable $\vec{\lambda}_1(s,x)$ plotted in (a).

Figure 23(a) plots $\zeta_\kappa(\sigma)$ and $h_{1,\kappa}(\sigma)$ in grey for fixed t, tracking the ends of $\vec{\kappa}$ and $\vec{\mathcal{R}}_\kappa$ respectively as surrogates for the points $x\vec{\kappa}$ and $x\vec{\mathcal{R}}_\kappa$ and serving to illustrate

$$\vec{h}_{1,\kappa-1}(s) + x\vec{\mathcal{R}}_\kappa + \vec{\zeta}_\tau(s) = \vec{\zeta}_{\kappa-1}(s) + x\vec{\kappa}. \tag{79}$$

The terms "bigger" and "smaller" are used to refer to the non-uniform enlargement and reduction of a pathway whilst accommodating differential scaling along that pathway. Figure 23(a) shows that for Euler's zeta when $\sigma < 1/2$ the pathway $P(h_{1,\kappa}(s))$ is "bigger" than $P(\zeta_\kappa(s))$, and when $\sigma > 1/2$ the pathway $P(h_{1,\kappa}(s))$ is "smaller" than $P(\zeta_\kappa(s))$. This difference in behaviour is important and is similar for $\eta_l(s)$ with $l \geq 2$; when $\sigma < 1/2$ a $P(h_{l,\kappa}(s))$ is "bigger" than its paired $P(\ell_{l,\kappa}(s))$, and when $\sigma > 1/2$ a $P(h_{l,\kappa}(s))$ is "smaller" than its paired $P(\ell_{l,\kappa}(s))$.

4.8. The pathway for Euler's zeta, $P(\zeta_n(s))$ is the Bauplan for $P(\eta_l(s))$

A *"Bauplan"* is a biological term for a collection of morphological features shared amongst members of a group. Euler's *zeta* is the *Bauplan* for

$P(\eta_l(s))$. We expect Euler's zeta $\zeta_\tau(\rho) \approx 0$ and for this to be more challenging at low values of t than at high values. However, even at the first zero $t_i = 14.1347...$, for which $\tau = 5$ and $\kappa = 2$, which we met in Figure 18(a) the challenge is well taken. This zero is shown again in Figure 24, with the 5 vectors of $P(\zeta_\tau(1/2 + it_1))$ in grey, $P(\zeta_\kappa(1/2 + it_1))$ in blue, $P(h_{1,\kappa}(s))$ in red and the line of reflection $\vec{\psi}_0$.

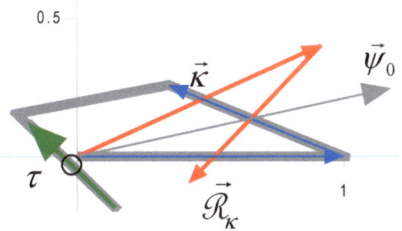

Figure 24. In grey, $P(\zeta_\tau(s))$ for $t = 14.1347$ and $\sigma = 1/2$ with its 5 $\vec{m}$; the fifth (in green) almost opposes the fourth. Here $\tau = 5$ and $\kappa = 2$. $P(\zeta_\kappa(s))$ in blue crosses $\vec{\psi}_0$. In red is $P(h_{1,\kappa}(s))$. The average of the τ vector and its predecessor lies in the black circle close to zero but not at zero.

As t_i rises through subsequent zeros, $P(\zeta_\tau(\rho_i))$ has the middle of its τ vector, or an appropriate average of neighbours, lying even closer to zero. Figure 25, below, compares $P(\zeta_\tau(s))$ with $P(\eta_l(s))$ showing pathways for $l = 2$ to $l = 5$; one for each corresponding vector in $P(\zeta_\tau(s))$.

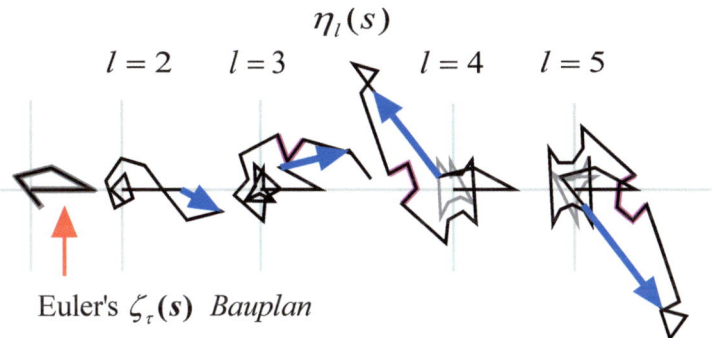

Figure 25. On the left the 5 vectors of $P(\zeta_\tau(s))$ with $t_i = 14.1347...$ and $\sigma = 1/2$. This *Bauplan*, $P(\zeta_\tau(s))$, is followed by the 4 pathways $P(\eta_l(s))$ for $l = 2$ to $l = 5$, each shown until convergence is evident. All 4 converge to zero. The blue arrows are the first l^{th} vectors in each series. If the first 5 vectors of each pathway are compared it can be seen that matched vectors share arguments, but in each there is a phase shift of π for the lth vector.

Even at the low value of t, used for Figure 25, near-regular polygons, with one side missing, are seen in the rudimentary P($\mathcal{R}_1$); a triangle ($l = 3$), a square ($l = 4$) and a pentagon ($l = 5$), are highlighted. When $l = 2$, at higher values of t the equivalent of the near-regular polygon with a missing side is a single vector between two nearly collinear negated vectors. Figure 25 illustrates the preservation of a *zero* on altering l. All pathways share $\vec{1}$ but necessarily *occupy* quite different regions of $\mathbb{R}^2$, and so when $s \notin \{\rho\}$ they converge at different points in $\mathbb{R}^2$. The curve traced by $\eta_l(\sigma)$ is a manifestation of that occupation. Remembering that $\sigma_\alpha < 1/2 < \sigma_\beta$ this mechanism precludes $\eta_l(\sigma_\alpha) = \eta_l(\sigma_\beta)$ for all l, whatever the value of t. For this reason, RH cannot fail irrespective of the value of t.

4.9. The pathway to convergence for Dirichlet's eta function ($l = 2$)

Figure 26 illustrates P($\eta_2(s)$), in blue, $s = \rho_{18961}$, a randomly chosen zero. The *proximal* pathway has in sequence, a set of $\vec{m}$ starting with $\vec{1}$ and proceeding with diminishing magnitudes $m^{-\sigma} > (m+1)^{-\sigma} > (m+2)^{-\sigma}$.... These *proximal* vectors do not form an obvious superstructure. False superstructures are possible if the reduced arguments $\Delta\theta$ (2.4. Preliminaries 2) form an "opportunistic sequence" (see Appendix A6) but this is unusual.

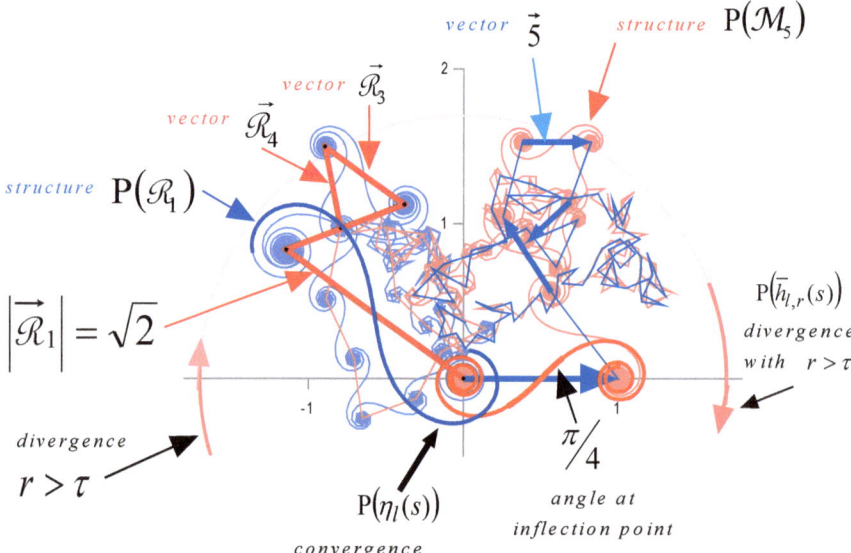

Figure 26. Two pathways are shown; **P($\eta_l(s)$)** in blue and **P($\overline{h}_{l,\tau}(s)$)** in red for $l = 2$, $t_i = 17224.28267$ and $\sigma = 1/2$. Both pathways start and end at 0. The $\vec{m}$ vectors $\vec{1}, \vec{3}, \vec{5}$ and $\vec{7}$ are shown in bold in blue. The $\vec{\mathcal{R}}_r$ (red) are plotted with $\vec{\mathcal{R}}_1$, $\vec{\mathcal{R}}_2, \vec{\mathcal{R}}_3$ and $\vec{\mathcal{R}}_4$ in bold with $\chi_l = \{2, 6, 10, 14 ...\}$. In pink, part of the encircling clockwise divergent pathway of P($\overline{h}_{l,r}(s)$) is indicated for an interval in $r > \tau$.

Once $\ln(m/(m-1)) - \ln((m+1)/m) < 2\pi$ sets of sequential $\vec{m}$ form the paired pseudo-spirals P($\mathcal{R}_r$). A finite sequence of P($\mathcal{R}_r$) runs as r falls, until $r = 1$, and these are also shown in blue in Figure 26. The last and largest pair of pseudo-spirals is P($\mathcal{R}_1$) whose *distal* diminishing spiral embraces convergence. The $|\vec{\mathcal{R}}_1| = \sqrt{l}$ when $\sigma = 1/2$.

The *principal-axis* of each P($\mathcal{R}_r$) has a relationship with an $\vec{\mathcal{R}}_r$ as shown in red in Figure 26. Importantly, the $\vec{\mathcal{R}}_r$ have an identity independent of the P($\mathcal{R}_r$) and are summated in the complimentary P($\bar{h}_2(s)$), such that when r is large the vector $\vec{\mathcal{R}}_r$ has an existence even when P($\mathcal{R}_r$) structures are not formed by P($\eta_2(s)$).

In Figure 26, P($\bar{h}_{l,r}(s)$) is shown in red alongside P($\eta_2(s)$). *Proximally* P($\bar{h}_{l,r}(s)$) has, in sequence, a finite set of vectors, and since $l = 2$ they have diminishing magnitudes $|\vec{\mathcal{R}}_r| > |\vec{\mathcal{R}}_{r+1}| > |\vec{\mathcal{R}}_{r+2}|$ ... . The series ends *distally* in paired pseudo-spirals P($\mathcal{M}_m$) before divergence. Divergence is indicated with a growing pink arc when $r > \tau$. Though covert, the symmetry between the arguments of $\vec{1}, \vec{3}, \vec{5}$ and $\vec{7}$ and the arguments of $\vec{\mathcal{R}}_1, \vec{\mathcal{R}}_2, \vec{\mathcal{R}}_3$ and $\vec{\mathcal{R}}_4$ is inescapable.

Figure 27 shows a rotated version of P($\bar{h}_{l,\tau}(s)$) with its first 24 vectors in bold. In black, the first 24 odd vectors of P($\eta_2(s)$) are placed in sequence, so omitting the alternating or negated even vectors. It is clear that for these vectors their arguments are mirror images and there is a line of reflection, the magnitudes however, are not equal.

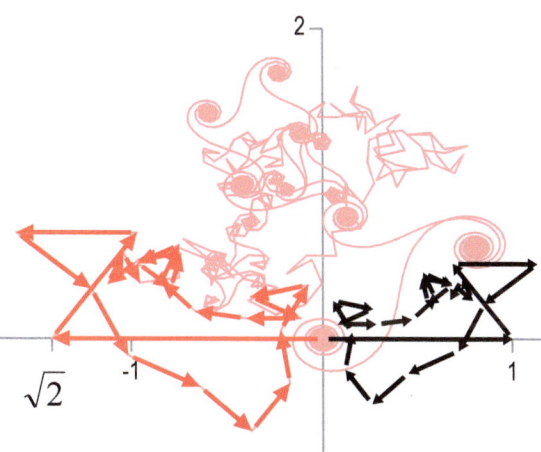

Figure 27.

This modification of Figure 26 shows a rotated P($\bar{h}_{l,\tau}(s)$) in red. The odd early vectors of **P($\eta_l(s)$)** are extracted as a novel series shown with black arrows. The mirror imaging of arguments is inescapable.

The symmetries in P($\eta_l(s)$), exemplified in Figures 26 and 27, can be extracted by allowing l to rise.

4.10. Symmetry-breaking in $\eta_{l,\infty}(s)$

We have $\eta_{l,n}(s)$ to κl terms as $\eta_{l,\kappa l}(s) = \zeta_{\kappa l}(s) - l^{(1-s)}\zeta_{\kappa}(s)$ and the vector $\vec{\mathcal{L}}_n = -l^{(1-s)}n^{-s}$ summated in $\ell_{l,\kappa}(s) = -l^{(1-s)}\sum_{n=1}^{\kappa} n^{-s} = \sum_{n=1}^{\kappa} \vec{\mathcal{L}}_n$ which together give us

$$\eta_{l,\kappa l}(s) = \zeta_{\kappa l}(s) + \ell_{l,\kappa}(s), \tag{80}$$

whose importance lies in $\zeta_{\kappa l}(\rho)$ being close to $\zeta_{\tau}(\rho) = 0$. We now have

$$\eta_{l,\infty}(s) \cong \zeta_{\kappa l}(s) + \ell_{l,\kappa}(s) - h_{l,\kappa}(s), \tag{81}$$

which can be refined by x to accommodate the intersection of the *kappa* vectors in *lambda*;

$$\vec{\lambda}_l(s,x) = \vec{\ell}_{l,\kappa-1}(s) + x\vec{\mathcal{L}}_\kappa - x\vec{\mathcal{R}}_\kappa - \vec{h}_{l,\kappa-1}(s), \tag{82}$$

with $\eta_{l,\infty}(s) = \zeta_{\kappa l}(s) + \lambda_l(s,x).$ \hfill (83)

This is visualised in $\mathbb{R}^2$ as $\vec{\lambda}_l(s,x)$ being the difference between the ends of $P(\vec{\ell}_{l,\kappa-1}(s) + x\vec{\mathcal{L}}_\kappa)$ and $P(\vec{h}_{l,\kappa-1}(s) + x\vec{\mathcal{R}}_\kappa)$ with both pathways starting at the origin.

We now move towards extracting the symmetries by allowing l to rise above *kappa*. The symmetry has to be appreciated first with $\sigma = 1/2$ and then we think about what happens when t and then σ change.

4.11. Setting $l > \kappa$: an example of a $P(\eta_l(s))$ when $l > \kappa$

The number of *proximal* vectors in $P(h_{l,n}(s))$ and $P(\ell_{l,n}(s))$, which have mirror image arguments about $\vec{\psi}$, counting from the origin, where we anchor the pathways, rises as l increases.

When $l > \kappa$, all sequential arguments between neighbouring vectors of the finite $P(h_{l,\kappa}(s))$ are mirror images of those arguments between the corresponding neighbouring vectors of the finite $P(\ell_{l,\kappa}(s))$.

Figure 28 (opposite on page 53) shows a $P(\eta_l(s))$, in green, with $l > \kappa$ using $P(\eta_5(1/2 + it_l))$ for $t_{33} = 107.1686$ where $\kappa = 4$. As expected, the 20th vector, $\vec{\kappa l}$, crosses $\vec{\psi}$. Figure 28 is shown in preparation for our appreciation of the changes illustrated in Figure 29 on page 54 and Figure 30 on page 55.

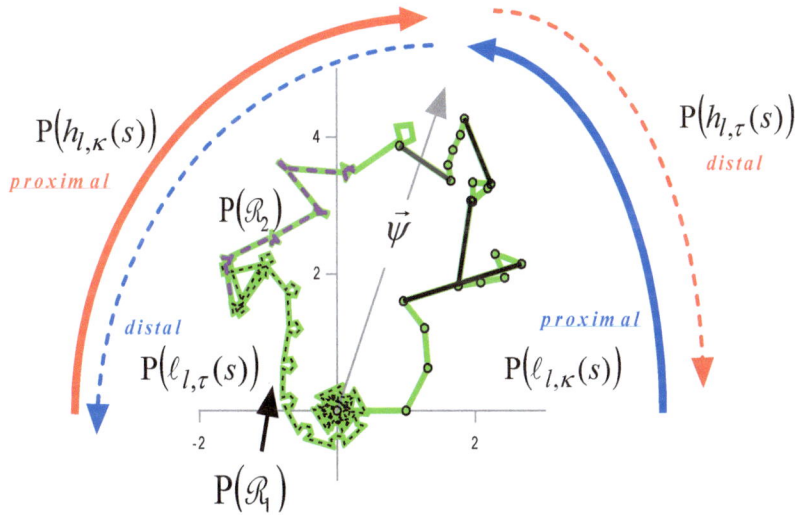

Figure 28. $P(\eta_5(1/4 + it_i))$ for $t_{33} = 107.1686$, in green, $\kappa l = 20$. The *proximal* part, $P(\eta_{5,\kappa l}(s))$ under the solid blue arrow will be tracked by $P(\ell_{5,\kappa}(s))$ to its *kappa* vector (not shown) and will be superimposed upon by the *distal* part of $P(\bar{h}_{5,\tau}(s))$ (not shown). The *distal* part of $P(\eta_5(s))$ under the dashed blue arrow will be tracked by the *distal* part of $P(\ell_{5,\tau}(s))$ and this will be superimposed upon by the *proximal* part of $P(h_{5,\kappa}(s))$. The vectors of $P(\eta_5(s))$ up to $\vec{\kappa l}$ have open dots at their extremities and the first 4 l^{th} vectors have a dark line joining those dots; these are the $\vec{\mathcal{L}}_n$ for $n = 1$ to $n = 4$.

In Figure 28 $P(\eta_5(s))$ has *proximal* and *distal* parts separated at an $\vec{m}$ where $m = \kappa l$ shown in black crossing $\vec{\psi}$. The *proximal* part under the solid blue arrow will be tracked by $P(\ell_{5,\kappa}(s))$ as far as $\vec{\psi}$ at its *kappa* vector $\vec{\mathcal{L}}_\kappa$. The *distal* part of $P(\eta_5(s))$ lies under the dark dashed blue arrow which will be tracked in the forward direction by $P(\ell_{5,\tau}(s))$ after the line of reflection $\vec{\psi}$ and after its *kappa* vector $\vec{\mathcal{L}}_\kappa$.

The $P(\mathcal{R}_1)$ is shown with a black dashed line over the green pathway. The function $\eta_5(s)$ has a weaker looking $P(\mathcal{R}_2)$ shown with a dashed purple line.

The next thing to appreciate is what happens when σ or t change and Figure 28 is the basis for Figure 29 on page 54 overleaf.

Figure 29(a) below, repeats plots for the 33rd zero showing $P(h_{5,\tau}(s))$ in red and $P(\ell_{5,\tau}(s))$ in blue. Alterations are made to σ in Figure 29(b) and to t in Figure 29(c).

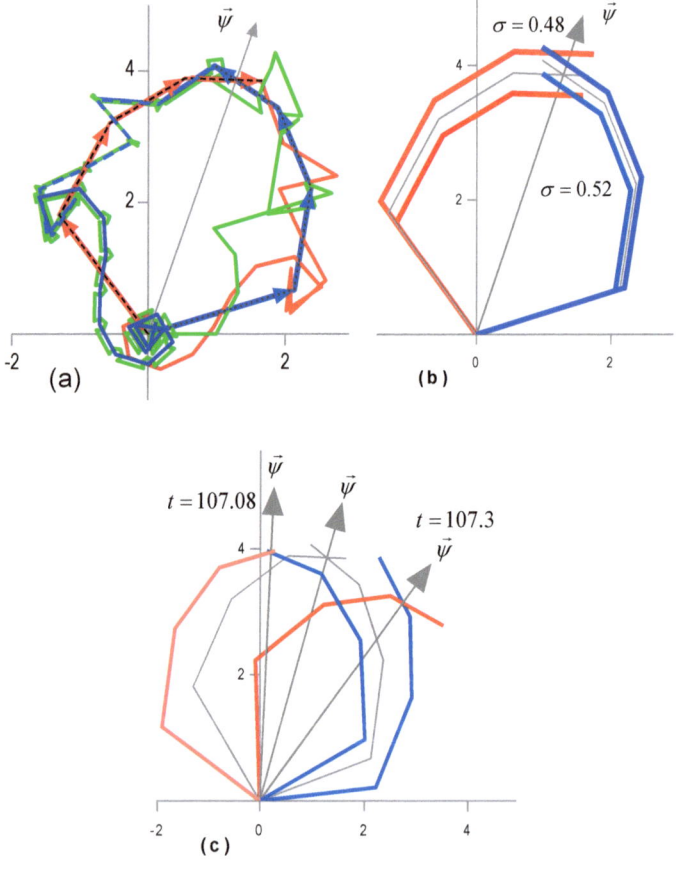

Figure 29. (a) $P(\eta_5(1/2 + it_i))$ in green for $t = 107.1686$. In blue $P(\ell_{5,\tau}(s))$ tracks $P(\eta_5(s))$ going forwards and in red is $P(h_{5,\tau}(s))$ the mirror image of $P(\ell_{5,\tau}(s))$. The 4 vectors up to and including $r = n = \kappa$ have arrow heads. The *kappa* vectors intersect on the line $\vec{\psi}$ in symmetry. (b) $P(\ell_{5,\kappa}(s))$ in blue and $P(h_{5,\kappa}(s))$ in red are shown with two domains when $\sigma \neq 1/2$. The *kappa* vectors cross, but intersections are not on $\vec{\psi}$ and a *zero* is precluded. (c) As in (b) but all with $\sigma = 1/2$ for two values of t, for which the *kappa* vectors just intersect, so meeting symmetry requirements, making this an interval in t which can host a *zero*.

In Figure 29(c) the interval in t over which x falls from 1 to 0 is roughly from 107.06 to 107.36 and Figure 30(a) below, shows there is no overlap of the vectors outside this interval.

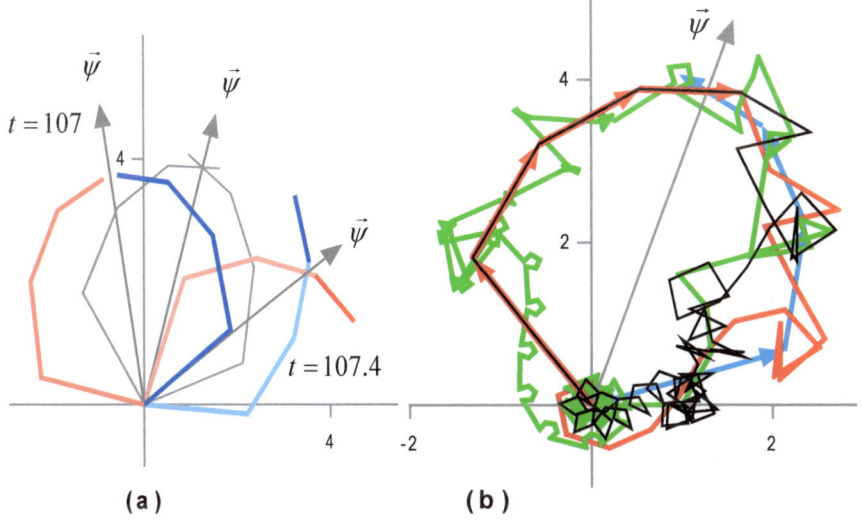

Figure 30. (a) $P(\ell_{5,\kappa}(s))$ in blue and $P(h_{5,\kappa}(s))$ in red are shown as in Figure 29(c) with a wider interval in t demonstrating no overlap of the *kappa* vectors. **(b)** In red $P(h_{5,\tau}(1/2+it))$ is shown which follows $P(\ell_{5,\kappa}(1/2+it))$ (light blue arrows), $P(\bar{h}_{5,\tau}(1/2+it))$ is shown in black. $P(\bar{h}_{5,\tau}(s))$ leaves $P(h_{5,\tau}(s))$ after the 5th vector and then retraces the *proximal* $P(\eta_{5,\kappa l}(s))$. Compare this with Figure 29(a) opposite on page 54.

The differences between $P(h_{l,\tau}(s))$ in red in Figure 29(a) opposite on page 54 and $P(\bar{h}_{l,\tau}(s))$ in black in Figure 30(b) above after the *kappa* vectors are shown for completeness. The *distal* $P(\bar{h}_{l,\tau}(s))$ overlies the *proximal* $P(\eta_l(\rho))$ whilst the *distal* $P(h_{l,\tau}(s))$ overlies the *proximal* $P(\ell_{l,n}(\rho))$.

4.12. An example of the paths of the mid-parts of the kappa vectors

Figure 31 uses three values of t with $\sigma = 1/2$ to illustrate the intersections of the *kappa* vectors over a short interval in t. The movement of the ends of the two pathways are shown with dashed lines over the same interval. These differ, with $P(h_{l,\kappa}(s))$ moving the furthest over the interval.

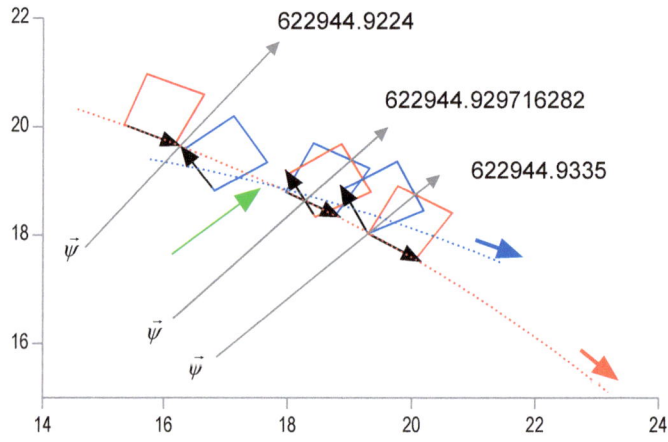

Figure 31. Three plots of the last 4 vectors of $P(h_{l,\kappa}(s))$ in red and $P(\ell_{l,\kappa}(s))$ in blue with *kappa* vectors shown as black arrows. The centre plot is for the 1,041,452nd zero at $t_i = 622944.9297$. The *kappa* vectors overlap throughout the short interval in t which ends just before 622944.9335. The crossing of the *kappa* vectors at the *zero* is not at their mid-points but in their tail sections. The arc in dashed red is $\left(h_{l,\kappa}(t) + h_{l,\kappa-1}(t)\right)/2$ and in dashed blue is $\left(\ell_{l,\kappa}(t) + \ell_{l,\kappa-1}(t)\right)/2$ over an interval in t from 622944.919 to 622944.944; these dashed lines cross at the green arrow ahead of $\vec{\psi}$ for the t_i consistent with the *zero* being associated with an intersection in the tails of the vectors and not at $x = 1/2$.

In Figure 31 for a fixed σ, over an interval in t, $h_{l,\kappa}(t)$ moves further than $\ell_{l,\kappa}(t)$ and *zeros* occur near the crossing of the arcs. The next *zero* occurs when $h_{l,\kappa}(s)$ catches $\ell_{l,\kappa}(s)$ once more.

4.13. The function $\lambda_l(s, x)$ identifies short intervals in t capable of hosting zeros

The series $\ell_{l,\kappa}(s)$, locates intervals in t which can host *zeros* on the *critical line*. Only one term of the *lambda* function is required; here the intersection between $\vec{\mathcal{L}}_\kappa$ and $\vec{\psi}$ is used. Since t specifies both $\vec{\psi}$ and κ and we let $l = \kappa + 1$ few calculations are required. The interval in t associated with the intersection of the *kappa* vectors gets narrower as t rises, and since κ and l grow slowly with t the computational burden rises slowly with t.

In this exercise t rises from 1,131,941.5 to 1,131,947.5 in increments of 0.0001. Values of t were noted when $\vec{\mathcal{L}}_\kappa$ intersected with $\vec{\psi}$ and those values immediately preceding intersection (Lower bound) and those immediately following intersection (Upper bound) appear in Table 2. The bounds are conservative as they lie outside of $x = 0$ and $x = 1$. In this interval $\kappa = 424$ and $l = 425$. In Table 2, if the column x contains [1→0] it indicates that the Lower bound is just outside an intersection of $x = 1$ and the Upper bound is just outside an intersection starting after $x = 0$.

Table 2.

$t = 1131940 + value\ in\ the\ table$

X	$\lambda(s,x) = 0$ Lower bound	t_i	$\lambda(s,x) = 0$ Upper bound	Δt Interval
[1→0]	1.8822	+ 1.8841	1.8887	0.0064
[0→1]	2.5649	+ 2.5708	2.5734	0.0084
[1→0]	2.9114	+ 2.9142	2.9208	0.0093
[0→1]	3.4800	+ 3.4852	3.4875	0.0074
[1→0]	3.9858	+ 3.9882	3.9938	0.0079
[0→1]	**4.4661**	**+ 4.4718**	**4.4743**	***0.0081***
[1→0]	4.9266	+ 4.9288	4.9337	0.0070
[0→1]	5.3665	+ 5.3696	5.3709	0.0043
[1→0]	6.1715	+ 6.1726	6.1753	0.0037
[0→1]	6.6373	+ 6.6413	6.6431	0.0057
[1→0]	7.0666	+ 7.0683	7.0723	0.0056

Intervals in t in which $\lambda_l(s,x)$ hosts 11 nontrivial zeros from t_i =1131941.8841 to t_i =1131947.0683. The figures are added to 1131940 to give t. The middle row shown in bold is the 2,000,000th zero.

Figure 32 (overleaf on page 58) spans the interval in t, covered by Table 2, plotting the magnitude of $\lambda_l(s,x)$ with $x = 1/2$. The 11 minima of the logarithm of the magnitudes fall very close to the 11 *zeros*, and the tightness of the upper and lower bounds is evident.

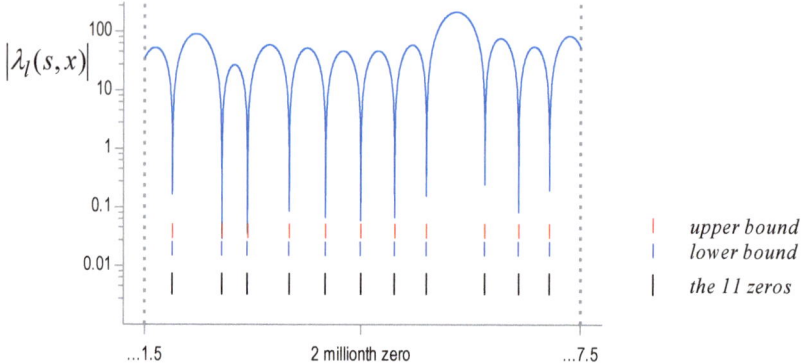

Figure 32. A plot of $\log|\lambda_l(s,x)|$ with $x = 1/2$; the upper and lower bounds indicate the limits of the intersection of the *kappa* vectors. The horizontal axis is t as in Table 2 on page 57.

The *zeros* in Figure 32 are precluded from having a Poisson-like distribution and "repel". The underlying mechanism creates a *zero* when the *kappa* vectors meet and the *zeros* repel as a consequence of the ends of $P(h_{l,\kappa}(s))$ and $P(\ell_{l,\kappa}(s))$ orbiting the origin.

4.14. Can an intersection of $\vec{\mathcal{L}}_\kappa(t)$ with $\vec{\psi}$ over an interval in t host more than one zero?

The exercise above was repeated for $\text{Im}(s)$ over the interval $t = 226$ to $t = 242$ with results shown in Table 3. In this interval $\kappa = 6$ and $l = 7$. Intersection equates to the interval in t for which $\lambda_l(s,x)$ could host a *zero*.

Table 3

$\lambda(s) = 0$ Lower bound	t_i	$\lambda(s) = 0$ Upper bound	Δt Interval
227.3533	227.4214	227.5012	0.1479
229.2312	229.3374	229.4338	0.2026
231.0702	**231.2502**	**232.1655**	**1.0953**
231.0702	**231.9872**	**232.1655**	**1.0953**
233.6252	233.6934	233.8016	0.1764
236.4073	236.5242	236.5967	0.1894
237.6827	237.7698	237.9533	0.2706
239.3638	239.5555	239.6391	0.2753
240.9733	241.0492	241.2516	0.2783

In Figure 33 (below), the magnitude of $\lambda_l(s,x)$ with $x = 1/2$ is shown with the upper and lower bounds being the start and end of the intersection of the *kappa* vectors. The locations of published *zeros* are also shown.

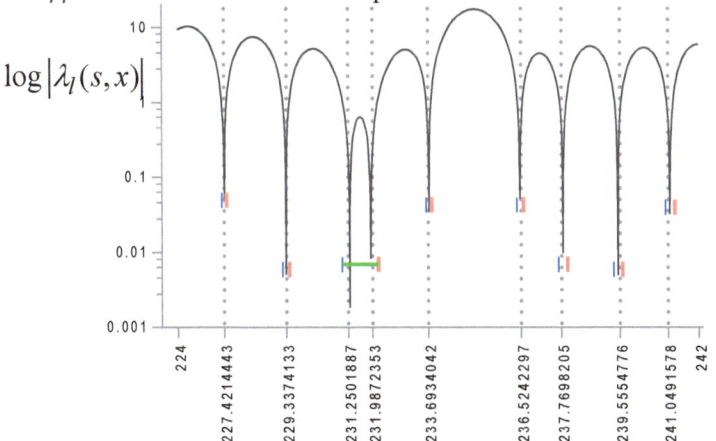

Figure 33. A plot of $\log_{10}|\lambda_l(s,x)|$ against t with $x = 1/2$. In red is shown the upper bound and in blue the lower bound. The dotted lines are the locations of the known *zeros*. One interval shown in green hosts two *zeros*.

In Figure 33 two *zeros* are hosted in the same short interval in t. In Figure 34 the paired pathways are illustrated at four values of t to show the underlying mechanism acting throughout this short interval.

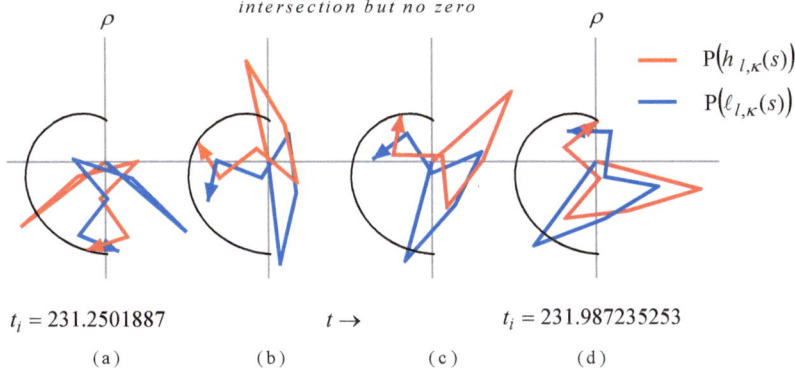

Figure 34. Four sequential plots, **(a)** with t_i as shown, **(b)** $t = 231.6$, **(c)** $t = 231.8$ and **(d)** with t_i as shown. Plots of $P(h_{l,\kappa}(s))$ in red and $P(\ell_{l,\kappa}(s))$ in blue over the interval between the two *zeros* during which the *kappa* vectors (arrows) are continually overlapping. The black arc in each figure is a plot of the movement of the tip of the $\vec{\mathcal{R}}_\kappa$ vector i.e. the function $h_{l,\kappa}(t)$ for $\sigma = 1/2$ over a slightly wider interval in t.

Figure 34 (page 59) shows how two *zeros* can be hosted in the same interval in t. This insensitivity of $\lambda_l(s,x)$ to detect all *zeros* does not weaken the symmetry arguments which prove RH. The same mechanism underlies some very short gaps between the *zeros*. Nonetheless, even if the gaps are small there is still "repulsion" of the *zeros*. Not all *zeros* require $\vec{h}_{l,\kappa}(s)$ to complete a full orbit before catching up once more with $\vec{\ell}_{l,\kappa}(s)$.

We consider further this short interval in t. In Figure 35 it can be seen that the magnitude of the angular velocity of $h_{l,\kappa}(s)$ in red is greater than that of $\ell_{l,\kappa}(s)$ in blue in the first part of the plot, but not the second part.

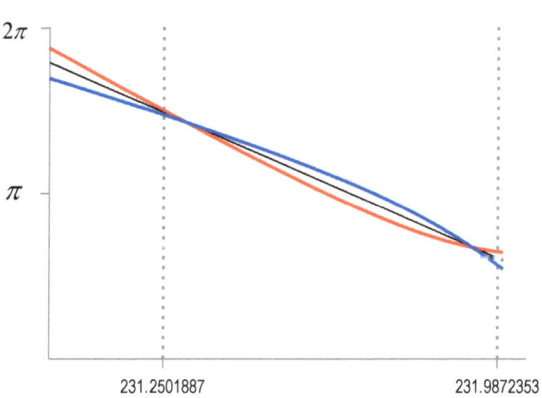

Figure 35. The arguments of the functions $h_{l,\kappa}(s)$ in red and $\ell_{l,\kappa}(s)$ in blue are shown with $\text{Arg}(\vec{\psi})$ in black over an interval in t. The arcs cross when the ends of the *kappa* vectors meet, equivalent to $x = 1$.

The crossing of the arcs represents the tip of the *kappa* vectors with the true *zeros* lying nearby with for $t_i = 231.250\ldots\ x = 0.591585\ldots$ and for $t_i = 231.987\ldots\ x = 0.599979$. Figure 35 complements Figure 34 on page 59.

Appendix A5 describes the empirical relationship between x, $\dot{\kappa}$ and the angle of intersection of the *kappa* vectors for the first 10,000 *zeros*. Appendix A5 illustrates a *zero* for which the intersection of the *kappa* vectors is nearly collinear (Figure A9 page 114) and one where the *kappa* vectors point in a similar direction (Figure A10 page 114).

4.15. A comparison of $\eta_l(t)$ and $\lambda_l(t)$ for fixed $\sigma = 1/2$

By Equation 83 (page 52) $\eta_l(s)$ and $\lambda_l(s,x)$ differ by $\zeta_{\kappa l}(s)$ and equate at the *zeros* when Euler's zeta vanishes. In Figure 36 (opposite on page 61) plots of $\eta_l(t)$ and $\lambda_l(t)$ for fixed $\sigma = 1/2$ can be seen to be similar, and the difference at one value of $t \notin \{t_i\}$ is illustrated to be $\zeta_{\kappa l}(s)$. In Figure 36 $l = 35$ and the first vector $\vec{\mathcal{R}}_1$, of the pathway $\text{P}(h_{l,\kappa})$ appears as a black arrow of magnitude $\sqrt{35}$. This vector is shown at the end of the reverse pathway indicated as $-\text{P}(h_{l,\kappa}) = \sum_{\kappa}^{1} \vec{\mathcal{R}}_r$, and shown in red and added onto the end of $\text{P}(\ell_{l,\kappa})$ rather than being anchored at the origin.

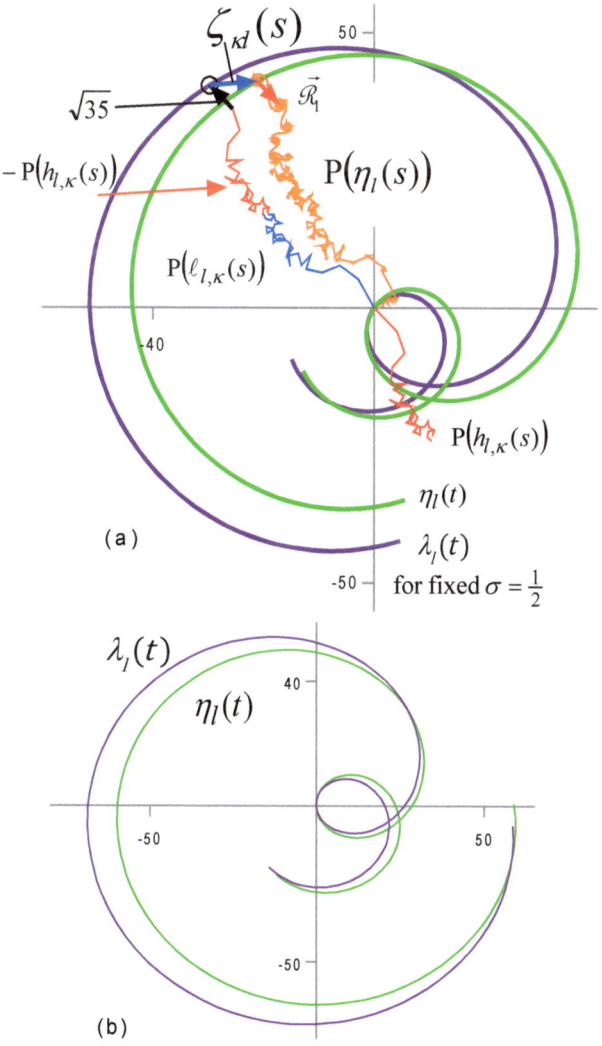

Figure 36. (a) Plots of $\eta_l(t)$ in green and $\lambda_l(t)$ in purple for $\sigma = 1/2$ over the interval $t = 7187.33269$ to $t = 7188.83269$ embracing the 6,914th zero with $\kappa = 34$, $l = 35$ and $x = 1/2$. With $s = 1/2 + i7187.70269$ the pathways are $P(\eta_l(s))$ in orange with $\vec{\mathcal{R}}_1$ a red arrow over $P(\mathcal{R}_1)$, $P(\ell_{l,\kappa}(s))$ in blue, $-P(h_{l,\kappa}(s))$ in red, and $\vec{\zeta}_{\kappa l}$ a blue arrow. (b) The same two curves but with $l = 67 = [\tau/\kappa]$.

In Figure 36(a) the arcs are similar but equate only at the *zero*. The other places where the arcs cross do not share the same value of t. Pathways are shown for $t = 7187.70269$ showing that the difference between $\lambda_l(s)$ and $\eta_l(s)$ is $\zeta_{\kappa l}(s)$ as indicated with the blue vector, $\vec{\zeta}_{\kappa l}(s)$.

The difference between $\lambda_l(s)$ and $\eta_l(s)$ at $t = 7187.70269 \notin \{t_i\}$ in Figure 36(a) is the blue vector $\vec{\zeta}_{\kappa l}(s)$ which is reproduced in Figure 37 below with its pathway $P(\vec{\zeta}_{\kappa l}(s))$.

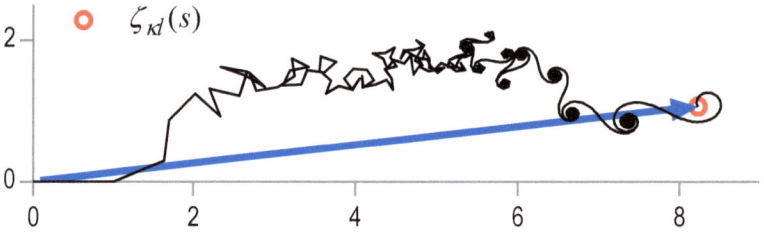

Figure 37. $P(\vec{\zeta}_{\kappa l}(s))$ from Figure 36(a) page 61 with $l = \kappa + 1$. The end point is not far from $\zeta_\tau(s)$. If t is large and $\vec{\zeta}_{\kappa l}(s)$ is based on $l = \kappa + 1$ then $\vec{\zeta}_{\kappa l}(s)$ will not completely disappear when $s \in \{\rho\}$.

Figure 37 shows $\vec{\zeta}_{\kappa l}(s)$ ending not far from $\zeta_\tau(s)$. If t is very large and $\vec{\zeta}_{\kappa l}(s)$ is based on $l = \kappa + 1$ then the term $\vec{\zeta}_{\kappa l}(s)$ will not completely disappear when $s \in \{\rho\}$. We can let l rise until $\kappa l \approx \tau$ when the final pseudo-convergence of Euler's *zeta* is reached. Strictly, we should let $l = [\tau/\kappa]$ to justify ignoring the partial Euler's *zeta* term when $\vec{\zeta}_{\kappa l}(\rho) \approx 0$, however, for illustrative purposes there is no need to be this precise.

4.16. Simultaneous zeros for $\eta_l(s)$ provide a metaphor for a ρ_u

Simultaneous zeros for $\eta_l(s)$ provide a metaphor for a ρ_u and help us dismiss any possibility that if $\{\rho_u\} \neq \emptyset$ then $\zeta_\tau(\rho_u) \neq 0$. It is argued that if $\{\rho_u\} \neq \emptyset$ then $\zeta_\tau(\rho_u) = 0$. A loop in $\eta_l(\sigma)$ for fixed t_u is required for RH to fail. In this section a loop in $\eta_l(\sigma)$ provides a metaphor for the pathways required for a ρ_u. We start with a loop in $\eta_l(\sigma)$ for fixed t, which allows a double-point $\eta_l(1/2 + it_i) = 0 \approx \eta_l(1 + it_i)$ when an appropriate l is chosen. We then imagine the loop as $\eta_l(s) = \vec{\zeta}_{\kappa l}(s) + \ell_{l,\kappa}(s) - h_{l,\kappa}(s)$ and think about how the associated pathways change as σ rises. In Figure 38(a) opposite, $\sigma = 1/2$ and only $P(\ell_{l,\kappa}(s))$ and $P(h_{l,\kappa}(s))$ are shown since $\vec{\zeta}_{\kappa l}(\rho_i) \approx 0$. In Figure 38(b) all three pathways are shown together with the loop in $\eta_l(\sigma)$ shown in green. This example uses the 1675th zero $t_i = 2170.787075$ for which $\kappa = 19$, and for which we set l at 20. This t_i was chosen since with l set at 20 we have $\ln(20)/(2\pi t_i) = 1035.00017$, which, being very close to an integer, gives a double-point close to zero. Importantly, we are concerned about the relationships between $P(h_{l,\kappa}(s))$, $P(\ell_{l,\kappa}(s))$ and $P(\vec{\zeta}_{\kappa l}(s))$ as they lie in $\mathbb{R}^2$ and how they change as σ rises.

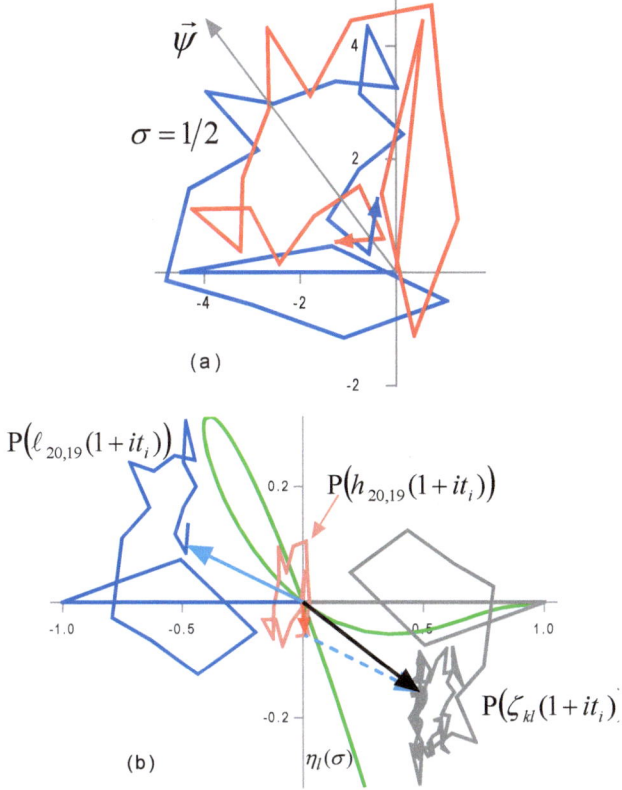

Figure 38. $P(h_{l,\kappa}(s))$ in red and $P(\ell_{l,\kappa}(s))$ in blue for $t_i = 2170.787075$ at **(a)** $\sigma = 1/2$ and **(b)** $\sigma = 1$, **(b)** also shows $P(\zeta_{\kappa l}(1+it_i))$ in grey. Plots are with $\kappa = 19$, $l = 20$ and $x = 0.254$. In **(b)** the black vector $\vec{\zeta}_{\kappa l}(1+it_i)$ equates to $\vec{h}_{l,\kappa}(1+it_i) - \vec{\ell}_{l,\kappa}(1+it_i)$, with vectors $\vec{h}_{l,\kappa}(s)$, shown with a red arrow and $-\vec{\ell}_{l,\kappa}(s)$, shown with a dashed blue arrow. The light blue arrow is $\vec{\ell}_{l,\kappa}(s)$. In green is the loop in $\eta_l(\sigma)$ with its double-point at zero.

Figure 38 illustrates $\eta_l(s) = \zeta_{\kappa l}(s) + \ell_{l,\kappa}(s) - h_{l,\kappa}(s) \approx 0$ for $\sigma = 1/2$ and $\sigma = 1$ for this carefully chosen s. The figure allows appreciation of how the trivial zero of the double-point, when $\sigma = 1$, comes about from the unique orientation or positioning of $P(h_{l,\kappa}(s))$ and $P(\ell_{l,\kappa}(s))$ in relation to $P(\zeta_{\kappa l}(s))$. This relationship would not exist if l were changed to a value l' that was not an integral power of l, i.e. $l' \neq l^a$ when $a \in \mathbb{N}_{>1}$.

It is for this reason that simultaneous *zeros* for some hypothetical t_u within the *critical strip*, could not be generated by a similar mechanism of collapse of a $P(\eta_l(s))$ over an interval embracing a σ_α and σ_β.

To illustrate the effect of a change in l the exercise was repeated with $l = 22$ and is shown in Figure 39.

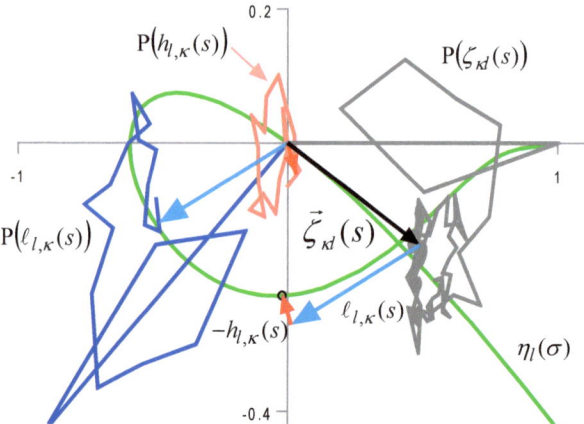

Figure 39. As Figure 38 but with $l = 22$. $P(\zeta_{\kappa l}(s))$ in grey is little changed but $P(\ell_{l,\kappa}(s))$ in blue and $P(h_{l,\kappa}(s))$ in pink are markedly different. The summation $\vec{\zeta}_{\kappa l}(s) + \vec{\ell}_{l,\kappa}(s) - \vec{h}_{l,\kappa}(s)$ reaches the green line of $\eta_l(\sigma)$ for $\sigma = 1$ at the back circle which is now not at zero. The black, blue and red arrows represent the same vectors as Figure 38. There is still a double-point in $\eta_l(\sigma)$ but not at zero. The green loop passes through zero for $\sigma = 1/2$ as expected.

At least one loop in $\eta_l(\sigma)$ is required with a double-point at zero for which $0 < \sigma_\alpha < 1/2 < \sigma_\beta < 1$ and for which the double-point is insensitive to changes in l if RH is false. There are two scenarios to consider. Firstly, if we accept that $\zeta_{\kappa l}(\rho_u) = 0$ we see that the symmetry breaking around changes in σ would prove RH. Secondly, if we entertain the idea that $\zeta_\tau(\rho_u) \neq 0$ we see that alterations in l bring about changes in $P(h_{l,\kappa}(s))$ and $P(\ell_{l,\kappa}(s))$ such that loops in $\eta_l(\sigma)$ cannot have fixed double-points, and this would prove RH.

5. The Derivatives: $\eta'_l(s)$, $h'_{l,r}(s)$, $\ell'_{l,n}(s)$ and $\zeta'_{l,\kappa l}(s)$

A line normal to $\vec{\psi}$ passing through the intersection of the *kappa* vectors is called the *"projection of the critical line"*. If RH is false, then given the Functional Equation, there must be an s for which $\eta'_l(s) = 0$ with $\text{Re}(s) < 1/2$. This is equivalent to saying that if $\text{Re}(s) > 1/2$ for all s which satisfy $\eta'_l(s) = 0$ then RH is true. This section shows why when $\text{Re}(s) < 1/2$ we cannot have $\eta'_l(s) = 0$. Consider a thought experiment ignoring $\zeta_{\kappa l}(s)$ and defining a *lambda* function $\lambda(s)$ without a subscript

$$\lambda(s) = \lambda_l(s, x) \text{ with } l = [\tau/\kappa] \text{ and } x = 1. \tag{84}$$

The behaviour of $\lambda(s)$ is now a surrogate for the behaviour of $\eta_l(s)$ and we ask about the zeros of $\lambda(s)$ and the zeros of $\lambda'(s)$. We call these "zeros" and not nontrivial "zeros" since they are not $\{\rho\}$.

When $\text{Re}(s) < 1/2$ and falls then $P(h_{l,\kappa}(s))$ grows larger more rapidly than $P(\ell_{l,\kappa}(s))$ grows. When $\text{Re}(s) > 1/2$ and rises then $P(h_{l,\kappa}(s))$ shrinks more rapidly than $P(\ell_{l,\kappa}(s))$. These behaviours explain why the *kappa* vectors separate one way on one side of the *projection of the critical line* and the other way on the other side of the *projection of the critical line* when the $\text{Re}(s)$ changes.

Changes in the magnitude of the individual vectors of pathways under changes in the $\text{Re}(s)$ are not uniform along the pathways but have a *proximo-distal* gradient. In $P(h_{l,\kappa}(s))$ the effects are greatest *proximally* and in $P(\ell_{l,\kappa}(s))$ the effects are more marked *distally*. These differences in behaviour account for why the separated *kappa* vectors lie on one side of $\vec{\psi}$ when $\sigma < 1/2$ and on the other side when $\sigma > 1/2$.

The derivatives counteract these two behavioural differences. It is immaterial that differentiation reverses the direction of every vector but importantly differentiation multiplies each $\vec{\mathcal{R}}_r$ by $\ln(v_1/r)$ and each $\vec{\mathcal{L}}_n$ by $\ln(ln)$, and so since $n \equiv r$ it can be appreciated how the behavioural differences outlined above are overcome only when $\sigma > 1/2$.

The zeros of $\lambda(s)$ lie on the *critical line* and the zeros of $\lambda'(s)$ to its right, but $\lambda(s)$ is not $\eta_l(s)$. Fortunately, we have no anxieties about the generalisation and can ignore $x = f(t)$, since this parameter is a refinement that breaks no symmetry but merely displaces the values of $\text{Im}(s)$ for the zeros of *lambda* a little from each neighbouring t_i. However, we do have anxieties over $\zeta_{\kappa l}(s)$ but these are easily dismissed. The term $\zeta_{\kappa l}(s)$ vanishes in the undifferentiated formulation at a nontrivial *zero* and can be comfortable ignored, but $\zeta'_{\kappa l}(s)$ does not vanish in $\eta'_l(s) = 0$ and it cannot be ignored. Nonetheless it can be shown that this causes no issues.

Differentiation of $|\vec{\mathcal{R}}_r| = \sqrt{l}\, v_1^{(1/2-\sigma)} r^{(\sigma-1)}$ with respect to σ gives

$$\left|\frac{\partial \vec{\mathcal{R}}_r}{\partial \sigma}\right| = \sqrt{l}\, v_1^{(1/2-\sigma)} r^{(\sigma-1)} \ln\left(\frac{v_1}{r}\right) = \ln\left(\frac{v_1}{r}\right) |\vec{\mathcal{R}}_r|,$$

or $\left|\dfrac{\partial \vec{\mathcal{R}}_r}{\partial \sigma}\right| = \sqrt{l}\, v_1^{(1/2-\sigma)} \left(\dfrac{q_{\bar{r}}}{2}\right)^{(\sigma-1)} \ln\left(\dfrac{2v_1}{q_{\bar{r}}}\right)$ with $q_{\bar{r}} \in \chi_l$ \hfill (85)

which are always positive since r is never greater than v_1 and the argument becomes

$$\arg\left(\frac{\partial \vec{\mathcal{R}}_r}{\partial \sigma}\right) = \frac{5\pi}{4} - t\ln\left(\frac{l}{r}\begin{bmatrix}v_1\\l\end{bmatrix}\right) = \arg(\vec{\mathcal{R}}_r) + \pi. \tag{86}$$

Since $\vec{\mathcal{L}}_n = -l^{(1-s)}n^{-s}$

$$\left|\frac{\partial \vec{\mathcal{L}}_n}{\partial \sigma}\right| = l^{(1-\sigma)}n^{-\sigma}\ln(ln) = \ln(ln)|\vec{\mathcal{L}}_n| \tag{87}$$

(with apologies for the repetitive symbols in the formulation $\ln(ln)|\vec{\mathcal{L}}_n|$) having an argument

$$\arg\left(\frac{\partial \vec{\mathcal{L}}_n}{\partial \sigma}\right) = \pi - t\ln(ln). \tag{88}$$

The series are then simply

$$\frac{\partial h_{l,\kappa}(s)}{\partial \sigma} = \sum_{r=1}^{\kappa} \frac{\partial \vec{\mathcal{R}}_r}{\partial \sigma} \tag{89}$$

and

$$\frac{\partial \ell_{l,\kappa}(s)}{\partial \sigma} = \sum_{n=1}^{\kappa} \frac{\partial \vec{\mathcal{L}}_n}{\partial \sigma}. \tag{90}$$

The differential of the partial Euler's *zeta* function is

$$\frac{\partial \zeta_{\kappa l}(s)}{\partial \sigma} = -\sum_{m=1}^{\kappa l} \ln(m)m^{-\sigma}(\cos(t\ln(m)) - i\sin(t\ln(m))). \tag{91}$$

Though Equation 81, on page 52, differentiates to

$$\frac{\partial \eta_l(s)}{\partial \sigma} = \frac{\partial \zeta_{\kappa l}(s)}{\partial \sigma} + \frac{\partial \ell_{l,\kappa}(s)}{\partial \sigma} - \frac{\partial h_{l,\kappa}(s)}{\partial \sigma}, \tag{92}$$

we find,

$$\frac{\partial \eta_l(s)}{\partial \sigma} = \frac{\partial \zeta_{\kappa l}(s)}{\partial \sigma} + \frac{\partial \lambda_l(s,x)}{\partial \sigma} \tag{93}$$

is less helpful than Equation 83 (page 52) since $\frac{\partial \zeta_{\kappa l}(s)}{\partial \sigma}$ does not vanish at the zeros of the differential.

5.1. A graphical analogy of Speiser's corollary: an example of $\eta'_l(s) = 0$

In preparation for looking at $\eta'_l(s) = 0$ Figure 40 shows three plots of $P(h_{l,\kappa}(s))$ in red and three of $P(\ell_{l,\kappa}(s))$ in blue. For this carefully chosen value of $t \notin \{t_i\}$ kappa = 34. Added on to the end of $P(\ell_{l,\kappa}(s))$ in blue, are shown the associated $P(\zeta_n(s))$ in green for the two plots that are not on the critical line.

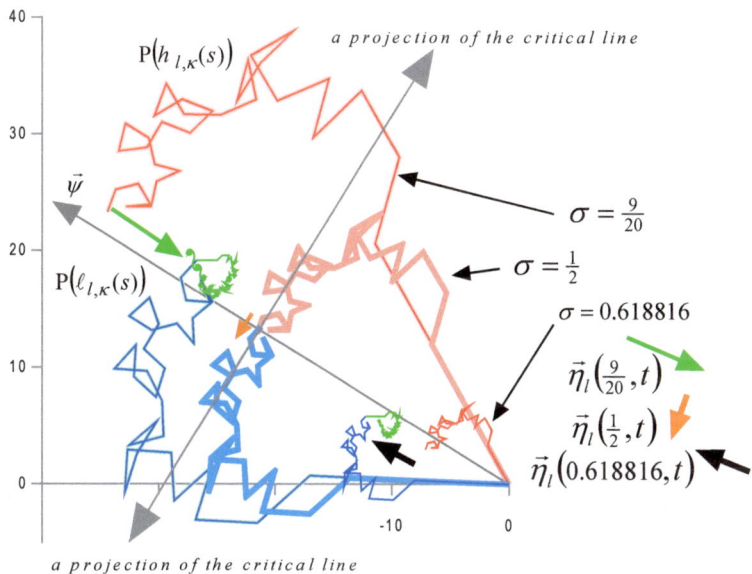

Figure 40. Three plots of $P(h_{l,\kappa}(s))$ in red and $P(\ell_{l,\kappa}(s))$ in blue for $t = 7188.332453 \notin t_i$ and real domains $\sigma = 9/20$, $\sigma = 1/2$ and $\sigma = 0.618816$. The three arrows approximate to the 3 vectors $\vec{\eta}_l(\sigma + it)$ shown in green, orange and black. The line of reflecton $\vec{\psi}$ is shown. The *projection of the critical line* is projected as a normal to $\vec{\psi}$, crossing it and passing through the $\sigma = 1/2$ pathway's *kappa* vectors.

In Figure 40 the index of negation is set at $l = 64$, with the justification for that value explained below. When $\sigma = 1/2$ the pathways are mirror images about the line of reflection $\vec{\psi}$. When $\sigma \neq 1/2$ the pathways separate in different, but predictable, ways above and below $\sigma = 1/2$. The breaking of symmetry either side of the *projection of the critical line* in Figure 40 is clear and understandable.

Differentiation provides a mechanism to "reverse" this breaking of symmetry but only when $\sigma > 1/2$. This is the graphical analogy of Speiser's corollary.

In Figure 40 on page 67, t was set at $t = 7188.332453$, which is a carefully chosen value slightly above $t_i = 7188.322595$. Near this t_i there is a trivial zero for $t = 1586\,\pi/\ln(2) = 7188.323185$ at $\sigma = 1$, and consequently there is a double-point in $\eta_l(\sigma)$ and a loop in $\eta_l(\sigma)$ over an interval in σ embracing $\sigma = 1/2$ and $\sigma = 1$. This loop has a minimum in $\frac{\partial \eta_l(s)}{\partial \sigma}$ near the apex of the loop (see Figure 41 opposite).

Minor adjustments to t from t_i soon establish the values of both t and σ for which $\eta'_l(s) = 0$ when $\frac{\partial \eta_l(s)}{\partial \sigma} = -i\frac{\partial \eta_l(s)}{\partial t} = 0$. We call these values t_c and σ_c. The adjustments narrow the loop in $\eta_l(\sigma)$ and move its double-point towards the loop's apex until it diassapears.

Unlike the set $\{t_i\}$, whose zeros $\eta_l(1/2 + it_i) = 0$ are independent of l, the zeros of $\eta'_l(s) = 0$ are dependent on l. Indeed, even though the zeros of $\eta_l(\sigma)$ at $\sigma = 1$ are invariant under integral powers of l, i.e. l^2, l^3, l^4, ... l^a,... with $a \in \mathbb{N}_{>1}$, the zeros of the differential $\eta'_l(s) = 0$ are sensitive to a. In Figure 40 $l = 64$.

For Figure 40, on page 67, having chosen $t = 1586\,\pi/\ln(2) = 7188.323185$, we have $\kappa = 34$ and an l was required satisfying $l > \kappa$ which was also an integral power of 2; the nearest exceeding 34 being $l = 64$.

Figure 41 (opposite on page 69) uses $l = 64$ and shows a loop in $\eta_l(\sigma)$ for $t_i = 7188.322595$ in blue, and a cycloid-like curve in $\eta_l(\sigma)$ for $t_c = 7188.332453$ over the interval $\sigma = 0.45$ to 3 in black.

Figure 41 also shows a cycloid-like curve in $\eta_l(t)$ with $\sigma_c = 0.618816$ in red. Passing through zero we have part of a prolate cycloid-like curve with $\sigma = 1/2$ in green and an arc in $\eta_l(t)$ with $\sigma = 1$ in orange.

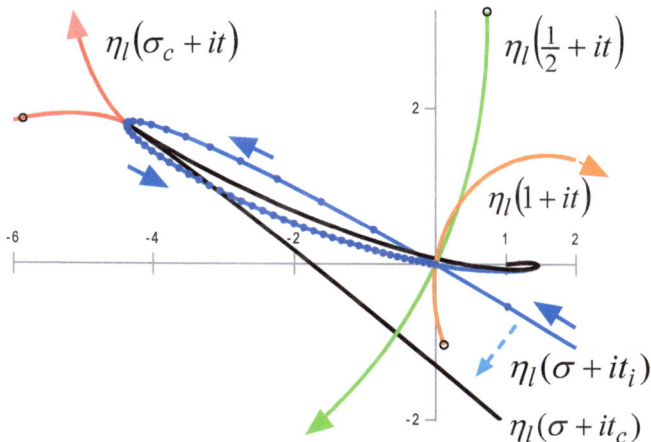

Figure 41. A loop in $\eta_{64}(\sigma)$ with t fixed at $t_i = 7188.322595$ is shown in blue with σ rising in the direction of the blue arrows. The loop passes through zero at $\sigma = 1/2$ and very near 0 at $\sigma = 1$. Equal increments in σ of 0.01 are indicated in the loop of $\eta_{64}(\sigma)$ with blue dots. For the black and red cycloid-like curves see the text.

Figure 41 shows how a small adjustment to t upwards from t_i narrows the blue loop, and moves the double-point from zero towards the apex of the loop. This produces the cycloid-like curve (in black) of $\eta_l(\sigma)$ at fixed $t = t_c$ (subscript c for cycloid) with its apex at $\eta_l(\sigma_c + it_c)$ where it meets the cycloid-like curve of $\eta_l(t)$ for fixed $\sigma = \sigma_c$ (in red).

Curves for $\eta_l(t)$ over short intervals embracing t_i are shown for $\sigma = 1/2$ in green, and for $\sigma = 1$ in orange, passing through the origin in roughly opposing directions; each cross $\eta_l(\sigma)$ at $\pi/2$. The loop in $\eta_l(\sigma)$, shown in blue, is with t fixed at $t_i \approx 2\pi k/\ln(64)$ (Roman $k \in \mathbb{N}_1$ and not Greek κ) so placing a double-point close to zero.

Figure 41 provides the elements for a mental image of the type of curve that must exist if RH were false. We imagine that $t = t_u$ rather than $t_i \approx \pi k/\ln(l)$ and the two values of σ at the double-point are σ_α and σ_β. The following would still apply: the differentials would relate as follows $\frac{\partial \eta_l(s)}{\partial t} = i\frac{\partial \eta_l(s)}{\partial \sigma}$, as seen with the two curves of $\eta_l(t)$, in green and orange, crossing the two regions of $\eta_l(\sigma)$ at $\pi/2$—note the directions indicated. The blue loop has a minimum in its differential at the apex, at a value lying closer to σ_α than σ_β; note the blue dots make rapid progress before slowing at the apex and then speeding up briefly, before finally slowing at high values of σ.

A minor adjustment to t allows the loop to shrink, and the two values of σ at the double-point approach one another until they meet at a value, we call, σ_c (c for cycloid) at which point we designate t as t_c. Two cycloid-like curves meet at $\eta_l(\sigma_c + it_c)$; in red $\eta_l(t)$ for $\sigma = \sigma_c$ over a short interval in t, and in black $\eta_l(\sigma)$ for $t = t_c$ over an interval in σ. Just as in Figure 41 page 69 where $(1 - \sigma_c) > (\sigma_c - 1/2)$, in our mental image $(\sigma_\beta - \sigma_c) > (\sigma_c - \sigma_\alpha)$, and since $\sigma_\beta = 1 - \sigma_\alpha$, we can say that if RH is false there must be a zero of the differential with $\sigma_c < 1/2$. This monograph shows that this inequality is impossible.

Figure 42 elaborates on the *kappa* vectors in Figure 40 (page 67) with $\sigma = 1/2$ and illustrates the partial differentials as vectors. Of relevance to the distribution of the size of the gaps between the *zeros* $(t_{i+1} - t_i)$ it is noted that the line $\vec{\psi}$ generally rotates faster than $\arg\left(\ell_{l,\kappa}(s)\right)$ but slower than $\arg\left(h_{l,\kappa}(s)\right)$.

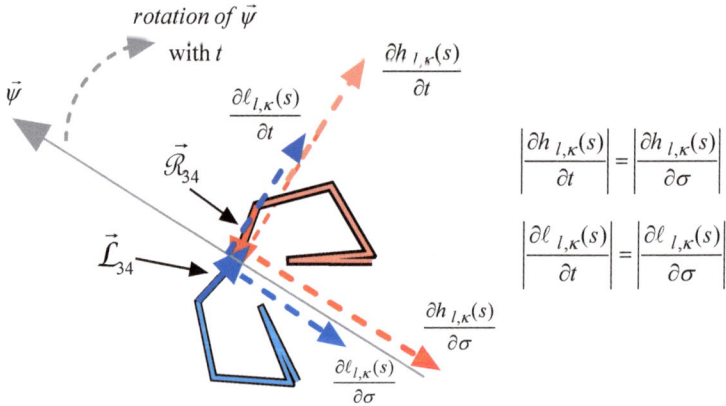

Figure 42. Detail from Figure 40 (page 67) showing the $30 - 34^{th}$ vectors of $P(h_{l,\kappa}(s))$ and $P(\ell_{l,\kappa}(s))$ as they cross $\vec{\psi}$. The ends of the vectors are the points $h_{l,\kappa}(s)$ and $\ell_{l,\kappa}(s)$ and their partial differentials are shown with their relationships; $\frac{\partial h_{l,\kappa}(s)}{\partial t} = i\frac{\partial h_{l,\kappa}(s)}{\partial \sigma}$ and $\frac{\partial \ell_{l,\kappa}(s)}{\partial t} = i\frac{\partial \ell_{l,\kappa}(s)}{\partial \sigma}$.

For values of t significantly removed from the vicinity of $t \in \{t_i\}$, the *kappa* vectors $\vec{\mathcal{R}}_\kappa$ and $\vec{\mathcal{L}}_\kappa$ of $P(h_{l,\kappa}(s))$ and $P(\ell_{l,\kappa}(s))$ do not intersect. In the vicinity of a *zero* the vectors may cross, and the crossing will be in symmetry if $\sigma = 1/2$. Figure 42 shows the meeting of $P(h_{l,\kappa}(s))$ and $P(\ell_{l,\kappa}(s))$, there is symmetry at $\sigma = 1/2$ but there will be no *distal* superposition when $t \notin \{t_i\}$. If the intersection of the *kappa* vectors were not in symmetry and we were to continue a little way further alonfg the pathways $P(h_{l,\tau}(s))$ and $P(\ell_{l,\tau}(s))$, by allowing r and n to rise above κ, we would soon find pathways separating and soon realise that $h_{l,\tau}(s) \neq \ell_{l,\tau}(s)$. This is why $x = f(\kappa)$ is the same for $x\vec{\mathcal{R}}_\kappa$ and $x\vec{\mathcal{L}}_\kappa$.

Figure 43 is to be considered in the context of Figure 40 (page 67). Figure 43 shows the zero of the differential. It can be seen that $P\left(\frac{\partial h_{l,\kappa}(s)}{\partial \sigma}\right)$ meets $P\left(\frac{\partial \eta_{l,\kappa l}(s)}{\partial \sigma}\right)$ with their final vectors intersecting.

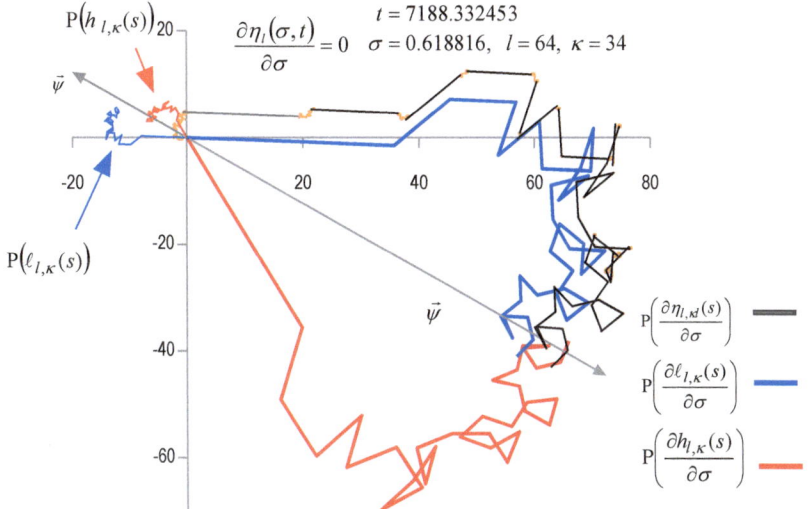

Figure 43. Near symmetry is restored when differentiation brings about a zero. $P(\ell_{l,\kappa}(s))$ in blue and $P(h_{l,\kappa}(s))$ in red (top left) have a value of $\sigma > 1/2$ as in Figure 40 (see page 67). Matched vectors have unequal magnitudes $|\vec{\mathcal{R}}_r| < |\vec{\mathcal{L}}_n|$, with $r = n$. However, on differentiation all vectors are reversed and enlarged. Enlargement is greatest for vectors in $P(h_{l,\kappa}(s))$ and so $\eta'_l(s) = 0$ becomes possible but only with $\sigma > 1/2$. $P\left(\frac{\partial \ell_{l,\kappa}(s)}{\partial \sigma}\right)$ is shown in blue, $P\left(\frac{\partial h_{l,\kappa}(s)}{\partial \sigma}\right)$ in red and $P\left(\frac{\partial \eta_{l,\kappa l}(s)}{\partial \sigma}\right)$ with every l^{th} vector in black and the intervening vectors in orange.

If at the end of the pathway $P\left(\frac{\partial \ell_{l,\kappa}(s)}{\partial \sigma}\right)$ we add the pathway $P\left(\frac{\partial \zeta_{\kappa l}(s)}{\partial \sigma}\right)$ we meet $P\left(\frac{\partial \eta_{l,\kappa l}(s)}{\partial \sigma}\right)$. This addition of $P\left(\frac{\partial \zeta_{\kappa l}(s)}{\partial \sigma}\right)$ is shown in Figure 44 overleaf on page 72.

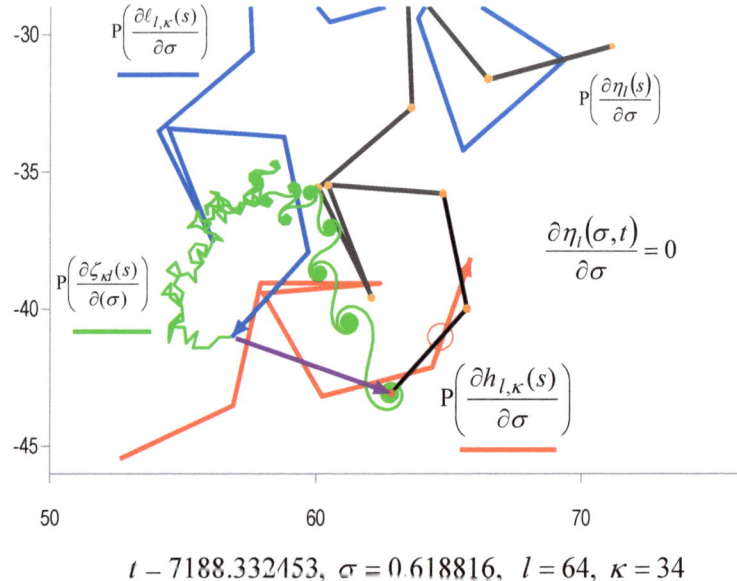

$t = 7188.332453$, $\sigma = 0.618816$, $l = 64$, $\kappa = 34$

Figure 44. With s as shown and $l = 64$ the function $\frac{\partial \eta_l(s)}{\partial \sigma} \approx 0$. The black line follows the final vectors near the end of $P\left(\frac{\partial \eta_{l,\kappa l}(s)}{\partial \sigma}\right)$, ending with $\vec{m}$ for $m = 34 \times 64 = 2176$. The orange dots lie at the end of every vector $\vec{m}$ and the lth vectors dominate the pathway. The same number of terms, 2176, is used for $\frac{\partial \zeta_{\kappa l}(s)}{\partial \sigma}$ (purple arrow) and its pathway $P\left(\frac{\partial \zeta_{\kappa l}(s)}{\partial \sigma}\right)$, which is shown in green; both start from the end of the *kappa* vector $\vec{\mathcal{L}}_\kappa$ terminating $P\left(\frac{\partial \ell_{l,\kappa}(s)}{\partial \sigma}\right)$, which is shown in blue. The $P(\zeta_{\kappa l}(s))$ ends at an $m = 2176$ in a red dot which is a little shy of $\tau = 2288$, but still very close to the final pseudo-convergence of Euler's zeta. The *kappa* vector of $P\left(\frac{\partial h_{l,\kappa}(s)}{\partial \sigma}\right)$ intersects with the κl vector of $\eta'_{l,\kappa l}(s)$ consistent with this being very close to a zero of the derivative.

In Figure 44 the $\frac{\partial \zeta_{\kappa l}(s)}{\partial \sigma}$ vector, in purple, runs between the ends of the *kappa* vectors but it could have been placed to incorporate the effect of x and it would then have ended in the red circle. Appendix A4 tabulates some calculations in relation to the magnitudes of the vectors for the differentials.

6. Why the *Zeros* Repel

This section describes how the paired finite vector series prevent the *zeros* from having a Poisson-like distribution. Accepting RH, we fix $\sigma = 1/2$ and require $l > \kappa$ and will use $l = \kappa + 1$ for calculations. We will assume that $x = f(\dot{\kappa})$ applies such that we can simplify notation by understanding the following functions free of subscripts

$$h(t) = h_{l,\kappa-1}(t) + x\vec{\mathcal{R}}_\kappa, \tag{94}$$

$$\ell(t) = \ell_{l,\kappa-1}(t) + x\vec{\mathcal{L}}_\kappa \tag{95}$$

$$\text{and } \lambda(t) = \ell(t) - h(t) \tag{96}$$

such that $h(t) = \ell(t)$ when $t \in \{t_i\}$ for an $x = f(\kappa)$ giving $\lambda(t_i) = 0$.

The line of reflection $\vec{\psi}$ is the line of symmetry for all paired vectors $\vec{\mathcal{L}}_n$ and $\vec{\mathcal{R}}_r$ when $n = r$ since $l > \kappa$ and it is also the line of reflection for the difference between all neighbouring vectors. It is the line of reflection for $P(h(t))$ and $P(\ell(t))$ since the paired vectors have equal magnitudes and the paired pathways set off from the origin. The line of reflection has two collinear rays meeting at the origin designated $\vec{\psi}_0$ and $\vec{\psi}_\pi$ with arguments ψ_0 and $\psi_\pi = \psi_0 + \pi$ respectively. At a *zero* the *kappa* vectors sometimes meet on $\vec{\psi}_0$ and sometimes on $\vec{\psi}_\pi$. We have no interest in the absolute magnitude of either ray. The formulation ψ_0 may represent the unreduced or the principal $[0, 2\pi)$ argument of the ray $\vec{\psi}_0$. Context will make clear if the principal (Arg) or if the unreduced (arg) form is being used. As t rises ψ_0 becomes progressively more negative, it is monotonic. The rays rotate clockwise ↻ and as t rises the magnitude of that rate of rotation increases.

The rate of change of ψ_0 (in its unreduced form) with t is

$$\dot{\psi} = \frac{d\psi_0}{dt} = \frac{d\psi_\pi}{dt}. \tag{97}$$

The first vector of $\ell(t)$ is $\vec{\mathcal{L}}_1$ which has an orientation of $\pi - t\ln(l)$, see Equation 36 page 21, and the orientation of $\vec{\mathcal{R}}_1$, the first vector of $h(t)$ is $\arg(\vec{\mathcal{R}}_1) = \pi/4 - t\ln(l[v_1/l]) = \pi/4 - t\ln(l[t/2\pi])$, see Equation 27 page 19. *Psi* is half the sum of these two orientations and the differential is half the sum of the two differentials;

$$\dot{\psi} = \frac{d\psi}{dt} = \frac{1}{2}\left(\frac{d}{dt}\arg(\vec{\mathcal{L}}_1) + \frac{d}{dt}\arg(\vec{\mathcal{R}}_1)\right). \tag{98}$$

It is clear that $\vec{\mathcal{L}}_1$ rotates clockwise ↻ less rapidly than $\vec{\mathcal{R}}_1$ since $t/2\pi > 1$.

However, if we turn our attention to the rotation of the vectors $\vec{h}(t)$ and $\vec{\ell}(t)$, these are not always clockwise since the dynamics of subsets within the pathways $P(h(t))$ and $P(\ell(t))$ can create cycloid-like behaviour. The rotation of the vectors of the pathways can generate curtate cycloid-like curves, true cycloid-like curves (with zero differentials) and prolate cycloid-like loops.

6.1. Cycloid-like behaviour in the pathways

Cycloid-like behaviour is easy to understand. The many vectors in a pathway, for example $P(\ell(t))$, could be summated in a novel order but the end point will follow the same arc $\ell(t)$. If the vectors representing the whole pathway, tracing a curve c, can be summated into two vectors $\vec{a}$ and $\vec{b}$, which are momentarily opposed to one another, and if those vectors have appropriate angular velocities and magnitudes, then cycloid-like behaviour follows, see Figure 45(a). Figure 45(a) uses only two vectors with fixed magnitudes and angular velocities and is necessarily symmetrical. However, except when *kappa* = 2, the summated vectors $\vec{a}$ and $\vec{b}$ can change both their magnitudes and angular velocities throughout a curtate cycloid-like curve or throughout a prolate loop and differ either side of the singularity in a true cycloid-like cusp. Consequently, the cycloid-like structures will not be symmetrical. The same mechanisms that apply to the dynamics of $P(h(t))$ and $P(\ell(t))$ apply to *eta*. Figure 45(b) illustrates the asymmetries either side of a true cycloid-like curve in part of a $P(\eta(s))$.

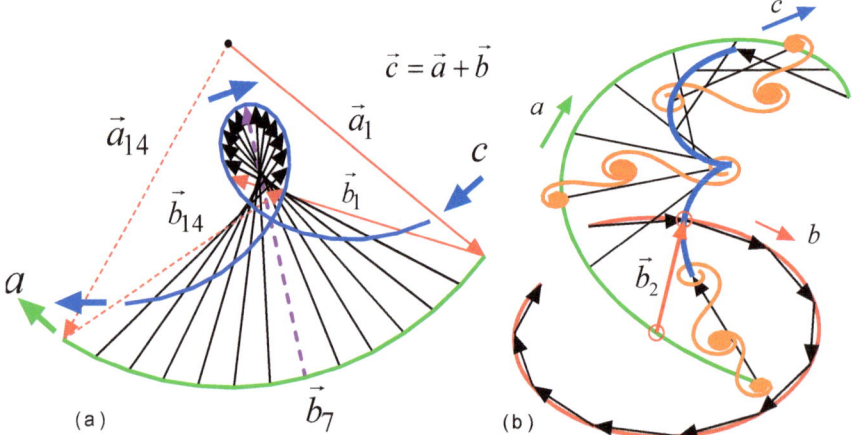

Figure 45. (a) The arc of $\vec{c}$ is the prolate cycloid-like curve c (blue), it is the sum of clockwise ↻ rotating $\vec{a}$ and $\vec{b}$. Arc a is in green. In red are $\vec{a}_1 + \vec{b}_1$ near the start of the loop and $\vec{a}_{14} + \vec{b}_{14}$ near its end. The dashed purple $\vec{b}_7$ opposes $\vec{a}_7$ (not shown). At the tip of the prolate loop in the slowest part of c ends $\vec{b}_7$. Curtate cycloid-like curves or a true cycloid-like curve would arise if $\vec{a}$ and $\vec{b}$ had different lengths or angular velocities. **(b)** A true cycloid-like curve showing summated vectors; asymmetry in c either side of its cusp is consequent upon changes over the interval in $\vec{a}$ and in $\vec{b}$ where $P(\mathcal{R}_2)$ is seen to changes its relationship with $P(\mathcal{R}_1)$.

In Figure 45(b) a true cycloid-like curve illustrates the asymmetry that follows from changes in the summated vectors $\vec{a}$ and $\vec{b}$. Figure 45(b) shows the arc c, for $\eta_{2,\tau}(1/2 + it)$ in blue, over an interval in t [1418.6, 1419.6], in green the arc a is $\eta_{2,116}(s)$, or $\vec{a} = \vec{\eta}_{2,116}(s)$. The final vectors summate to give $\vec{b}$. The interval in t is broken into 10 parts with 11 $\vec{b}$ vectors (10 of them indicted in black) advancing from arc a to arc c. The first and last have arrow heads. The second vector b_2 is in red and were it to keep its origin fixed it would trace out the arc b. Summated components are slightly different either side of the cusp. Vector $\vec{b}$ is composed of P($\mathcal{R}_1$) and P($\mathcal{R}_2$) and is shown with its many $\vec{m}$ in orange.

We know that $\ell(t)$ and $h(t)$ will have clockwise ↻ arcs with the occasional curtate cycloid-like curve, true cycloid-like cusp or prolate cycloid-like loop. It is the interaction of these arcs which give rise to the zeros. The gaps between the zeros arise as a consequence of these mechanisms which necessarily preclude Poisson-like randomness.

6.2. The meeting of the arcs of $\ell(t)$ and $h(t)$

In general, we expect $\left|\frac{d}{dt}(\vec{\ell}(t))\right| < \left|\frac{d}{dt}(\vec{h}(t))\right|$ and so, if t were time, the ends of the pathways would move at different speeds in $\mathbb{R}^2$. The point $h(t)$ would generally move faster than the point $\ell(t)$. The intricacies follow from the summation of the rotations of the vectors that constitute the pathways. As shown above the mechanics allow occasional cycloid-like curves, with instantaneously zero velocities and closely related curtate cycloid-like curves. Prolate cycloid-like curves form loops which fail to embrace the origin. Such loops are more common in $\ell(t)$ than in $h(t)$ at values of $t < 10^6$ (unpublished observation). If this difference were reversed and such loops were more common in $h(t)$ than in $\ell(t)$ at high values of t the arguments explaining the repulsion of the zeros that follow would still be valid.

The motions of the ends of the pathways are consequent upon their vector constructions. The link is prescribed by *Axiom 2* and the options for the behaviours of the ends of the pathways are limited to a clockwise rotation with occasional cycloid-like behaviour having a curtate, true, or prolate like nature. For example, as a consequence the $\arg(\vec{\ell}(t))$ may momentarily rise as the end of the pathway, when viewed from the origin, appears to move anticlockwise ↺. Both pathways progress in this way and zeros of zeta occur only when P($h(t)$) and P($\ell(t)$), meet at the same value of t.

An $x = f(\dot{\kappa})$ can oscillate such that $\lambda(\rho) = \zeta(\rho) = 0$ and it can be seen that a *zero* cannot lie too close to its immediate neighbour since after a *zero* the $\arg(\vec{h}(t))$ has to advance sufficiently to catch up with $\arg(\vec{\ell}(t))$ once more. A simple oscillating $x = f(\dot{\kappa})$ was used for the illustrations. The function is not sophisticated and is based on the first two terms of the Taylor series for the *sine* function which crudely resembles the relationship we desire (see Appendix A5 page 111–114).

The functions $\ell(t)$ and $h(t)$ are like runners in a race in $\mathbb{R}^2$ orbiting the origin but always on personal clockwise arcs. Viewed from the origin the runners generally move clockwise ↻, but occasionally one may pursue a small clockwise loop which excludes the origin. In this instance, the runner briefly moves anticlockwise ↺ when viewed from the origin. The $h(t)$ runner is generally faster than the $\ell(t)$ runner ($\vec{\mathcal{L}}_1$ rotates clockwise less rapidly than $\vec{\mathcal{R}}_1$). Every time $h(t)$ meets $\ell(t)$ there is a *zero*, signalled with runners exchanging a baton. Exchange requires runners to coincide in $\mathbb{R}^2$ at the same time and so if the faster runner laps the slower at a different distance from the origin there can be no *zero* even if the reduced arguments equate, $\text{Arg}(\vec{h}(t)) = \text{Arg}(\vec{\ell}(t))$.

The *zeros* cannot have a Poisson distribution because once $h(t)$ and $\ell(t)$ have met and the baton exchanged, $h(t)$ passes $\ell(t)$ and they cannot meet again until t has increased sufficiently to allow $h(t)$ to catch up with $\ell(t)$ once more.

There are three common ways that $h(t)$ can meet $\ell(t)$ again after meeting at a *zero*.

    a)   If the orbits are simple then $h(t)$ completes a little more than a whole number of laps before catching $\ell(t)$ again. This results in repulsion of the *zeros*.

    b)   If $\ell(t)$ takes a shorter radius arc than $h(t)$ its angular displacement can move ahead, even though it is moving slower. However, the faster $h(t)$ may now catch up with $\ell(t)$ before completing a lap, see Figure 46 opposite on page 77. Even though a whole lap is not completed there is still repulsion of the *zeros*. Rarely an extremely short interval between *zeros* is possible if the functions rotate at similar rates on similar arcs. Nonetheless, the faster function will cross the slower function creating the first *zero*, take a longer arc and then re-cross the slower arc to create the second *zero*. However close this pair lie, repulsion is still inherent and the multitude of rotating vectors in P($h(t)$) and P($\ell(t)$) cannot maintain this for long.

c) If $\ell(t)$ generates a prolate cycloid-like loop that does not encircle the origin it can reverse its angular progression in $\mathbb{R}^2$ as seen from the origin such that $h(t)$ catches $\ell(t)$ before completing a full rotation after the last *zero*. Once more the gap between *zeros* may be small but there is repulsion of the *zeros*.

Examples of intersecting arcs and the associated gaps between the *zeros* follow. We start with an example of (b) in Figure 46.

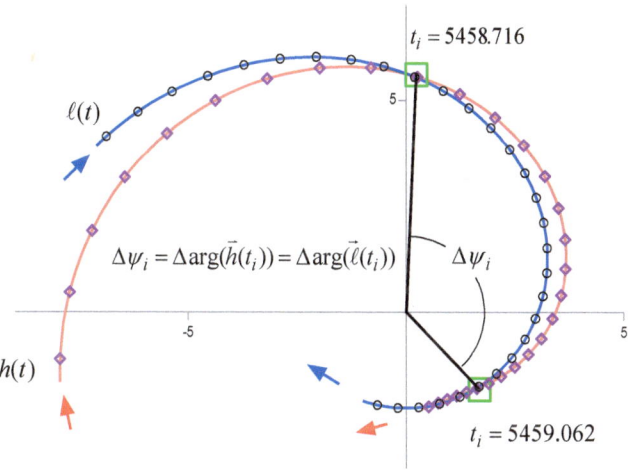

Figure 46. Two *zeros* $t_i = 5458.716...$ and $t_i = 5459.062...$ in an interval in t from 5458.5 to 5459.2, with $\ell(t)$ in blue and $h(t)$ in red equating in the green squares. Equal intervals in t are 1/15 of the inter-*zero* gap and are indicated on the arcs. After the first *zero*, the faster $h(t)$ takes the longer arc and $\ell(t)$ the shorter such that they meet again at the second *zero* before completing a rotation. This mechanism can place *zeros* very close together whilst maintaining repulsion of the *zeros*.

Whilst considering the examples which follow it is interesting to compare the clockwise progression in the unreduced argument of ψ_0 with the clockwise progression of the unreduced arguments $\arg(h(t))$ and $\arg(\ell(t))$. These are of interest when considering the statistical distribution of different measures for the gaps between the *zeros*.

6.3. A comparison of the rotations of $\vec{\psi}$, $\vec{h}(t)$ and $\vec{\ell}(t)$

For this illustration a short interval in t from $t = 1{,}126{,}903.423$ to $t = 1{,}126{,}905.0$, at the start of the region belonging to $\kappa = 424$ and embracing 4 *zeros* a, b, c and d, is shown in Figure 47. The curves are plotted at intervals in t of 0.001, as detailed in Table 4. Important features to appreciate include the two prolate cycloid-like loops in $\ell(t)$ in blue. These are clockwise loops but since they do not embrace the origin the $\arg\left(\vec{\ell}(t)\right)$ experiences a brief anticlockwise interlude as t rises.

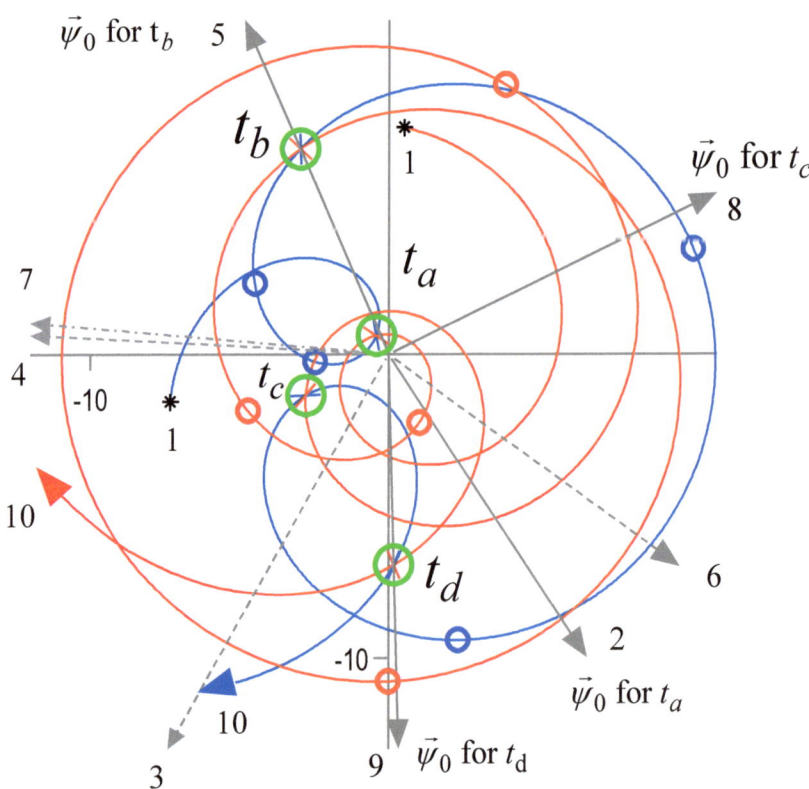

Figure 47. The mechanism underlying the creation of 4 *zeros*, t_a, t_b, t_c and t_d. The *zeros* are shown in green circles where the functions $\ell(t)$ in blue and $h(t)$ in red equate. The *kappa* vectors $\vec{\mathcal{L}}_\kappa$ and $\vec{\mathcal{R}}_\kappa$ are shown at each *zero* (in the green circles). For each *zero* only one ray, representing the line of reflection $\vec{\psi}_0$, is shown. For two *zeros*, t_b and t_d, the rays shown meet the *kappa* vectors; for the two *zeros* t_a and t_c the rays do not meet the *kappa* vectors as these *zeros* are associated with the $\vec{\psi}_\pi$ rays, which are not illustrated.

In Figure 47 opposite on page 78, the clockwise arcs of $\ell(t)$ in blue, and $h(t)$ in red start with a $*$ at position 1 and end with an arrow head at position 10. When viewed from the origin, the $\arg(\vec{\ell}(t))$ falls, except during the two loops that fail to embrace the origin, where momentarily it increases (the rotation appears to be anticlockwise). The red and blue circles indicate positions 3, 4, 6 and 7 in order along the arcs and are detailed in Table 4; the red and blue circles are reflections of one another either side of their associated $\vec{\psi}_\pi$, which are shown with numbered grey dashed lines 3, 4, 6 and 7.

Table 4. A guide to Figure 47 detailing each value of t illustrated.

Labels		t	Line of reflection	Intervals between zeros		
				$\Delta\psi_i/2\pi$	$\frac{\Delta\arg(\ell(t_i))}{2\pi}$	$\frac{\Delta\arg(h(t_i))}{2\pi}$
1	$*$ start	$t = 1126903.453$				
2	a	$t_a = t_i = 1126903.7$	2 $\vec{\psi}_0$			
3		$t = 1126903.826$	3	-0.528	-0.028	-1.028
4		$t = 1126903.917$	4			
5	b	$t_b = t_i = 1126904.00$	5 $\vec{\psi}_0$			
6		$t = 1126904.224$	6	-1.244	-0.744	-1.744
7		$t = 1126904.439$	7			
8	c	$t_c = t_i = 1126904.65$	8 $\vec{\psi}_0$			
				-0.3174	+0.18244	-0.81743
9	d	$t_d = t_i = 1126904.8$	9 $\vec{\psi}_0$			
10	↘✓	$t = 1126904.889$				

In Figure 47 in blue is $\ell(t)$ and in red $h(t)$ over an interval in t embracing four t_i as detailed in Table 4. Each arc passes through the four zeros labeled t_a, t_b, t_c and t_d where $\vec{\mathcal{L}}_\kappa$ and $\vec{\mathcal{R}}_\kappa$ intersect and meet their specific ray of $\vec{\psi}$. Detail from the centre of Figure 47 is reproduced in Figure 48 overleaf on page 80 to show the zeros t_a and t_c. This "t_c" has a c as an index and does not indicate a point on a cycoid-like curve.

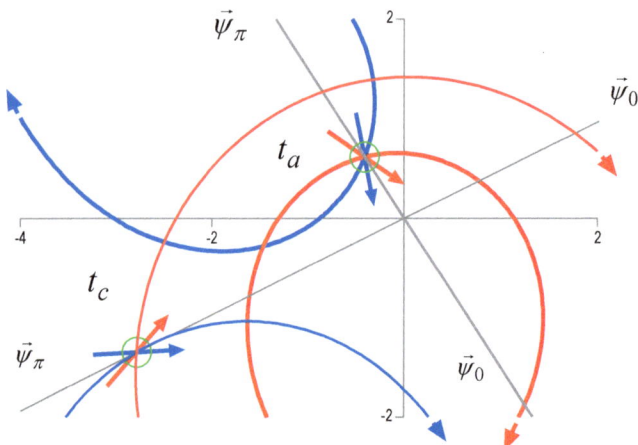

Figure 48. Detail from Figure 47, showing the intersection of the *kappa* vectors for t_a and t_c. The line of reflection $\vec{\psi}_0$ has an argument $\psi_0 = -\left(\frac{t}{2}\right)\ln\left(l^2\begin{bmatrix}v_1\\i\end{bmatrix}\right) - \frac{3\pi}{0}$ which when $l = \kappa + 1$ approximates to $-t\ln\left(\left[\frac{t}{2\pi}\right]\right) - \frac{3\pi}{8}$.

The reduced arguments $\mathrm{Arg}(\vec{\ell}(t))$ in blue and $\mathrm{Arg}(\vec{h}(t))$ in red from Figures 47 and 48 are plotted in Figure 49 and relate to Table 4.

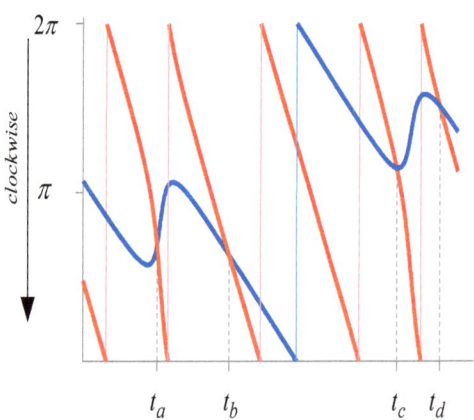

Figure 49. $\mathrm{Arg}(\vec{\ell}(t))$ in blue and $\mathrm{Arg}(\vec{h}(t))$ in red plotted against t following Figure 47 and Table 4.

Brief prolate loops of $\vec{\ell}(t)$ allow $h(t)$ to catch $\ell(t)$ swiftly; this places t_a close to t_b and t_c close to t_d.

Figure 49 sees $\vec{h}(t)$ complete more than 4 rotations whilst $\vec{\ell}(t)$ completes less than one. *Zeros* occur when both arguments equate and magnitudes equate. Here the magnitudes equate and curves cross at t_a, t_b, t_c and t_d. The rate of clockwise rotation of $\vec{h}(t)$ exceeds that of $\vec{\ell}(t)$, and $h(t)$ meets $\ell(t)$ at t_d only shortly after t_c without having to complete a rotation because of the second prolate loop in $\ell(t)$. Figure 50 (opposite on page 81) was produced by adding the rotations of $\vec{\psi}_0$ and $\vec{\psi}_\pi$ to a copy of Figure 49.

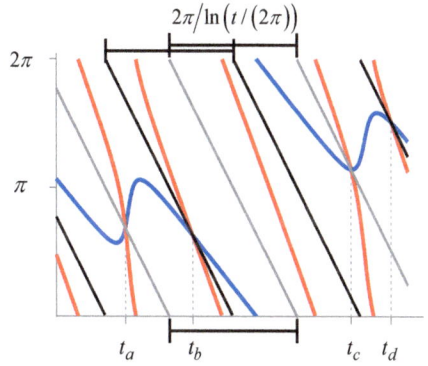

Figure 50. Arg($\vec{\ell}(t)$) in blue and Arg($\vec{h}(t)$) in red as in Figure 49 with ψ_0 in black and ψ_π in grey. The "wavelength" of ψ_0 is very close to $2\pi/\ln(t/(2\pi))$ as shown with the bar ⊢⊣. The *kappa* vectors intersect with either $\vec{\psi}_0$ or $\vec{\psi}_\pi$.

The "wavelength" in t, over which $\vec{\psi}_0$ completes a full rotation, is very close to $2\pi/\ln(t/(2\pi))$. This follows from Equation 37 (page 22) when $l = \kappa + 1$ and this heads to zero as t rises. The average gap between the *zeros* falls as t rises whilst the average rotation between ψ_0 for neighbouring *zeros* heads towards 2π with the orbiting mechanism continuing to preclude a Poisson-like distribution.

6.4. An illustration of a very small gap between zeros

In Figure 51 the interval in t from 663318 to 663320 is shown. This embraces three *zeros*, $t_a = 663318.5083$, then $t_b = 663318.5113$ which follows after a very small gap, less than 0.00296, followed by $t_c = 663319.8415$ with a larger gap exceeding 1.330.

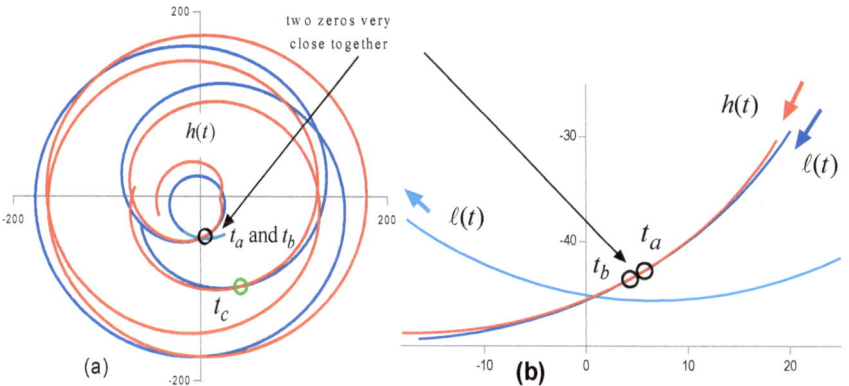

Figure 51. In **(a)** in red $h(t)$ and in blue $\ell(t)$ from $t = 663318$ to $t = 663320$ with three *zeros* (the start is shown in light blue). In **(b)** is shown the shorter internal interval 663318.47 to 663318.55 embracing the first two *zeros*. After the two very close *zeros* the function $h(t)$ and $\vec{\psi}$ both complete -2.949 rotations whilst $\ell(t)$ completes only -1.949 rotations before meeting at the third *zero* in the green circle.

Figure 51 (page 81) shows an example in which $h(t)$ and $\ell(t)$ generate two arcs which nearly superimpose and equate at two crossing places in close proximity. This is an extreme instance of that shown in Figure 46 on page 77. Importantly, pathway dynamics preclude such a mechanism from continuing to place *zeros* in such close proximity which would be required if the mechanics were to mimic a truly random Poisson-like distribution.

6.5. An illustration of a prolate cycloid-like loop in $h(t)$

The interval in t, in Figure 52(a), has 4 *zeros* and has, what is not commonly seen in the first 10^6 *zeros*, a prolate loop in $h(t)$ which does not encircle the origin. Figure 52(a) also has a loop containing t_b, and a loop which passes remarkably close to zero containing t_d. Figure 52(b) has the reduced arguments of the functions and the reduced form of ψ_0 in black.

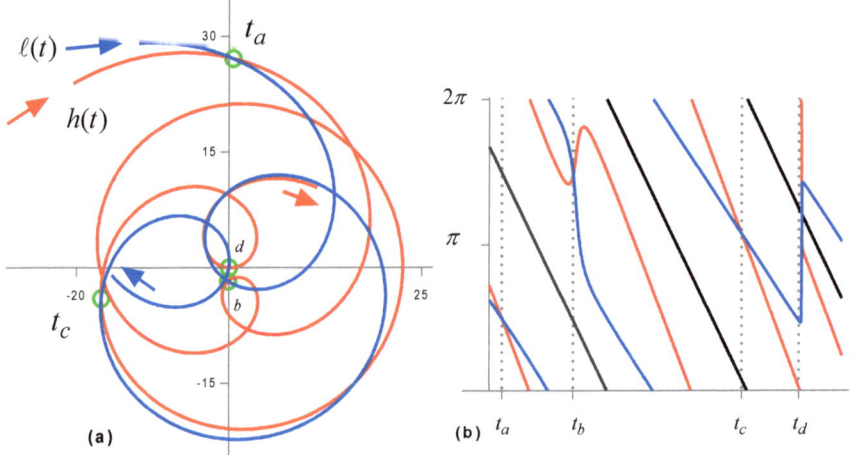

Figure 52. (a) In red $h(t)$ and in blue $\ell(t)$ from $t = 1130797.79$ to 1130799.1 with four *zeros*, $t_a = 1130797.837$, $t_b = 1130798.099$, $t_c = 1130798.724$ and $t_d = 1130798.948$ in green circles. Note the prolate cycloid-like loop in $h(t)$ which embraces t_b and the loops in $h(t)$ and $\ell(t)$ which pass remarkably close to zero. (b) The arguments are $\mathbf{Arg}(\vec{h}(t))$ in red, $\mathbf{Arg}(\vec{\ell}(t))$ in blue and ψ_0 in black.

Figure 52(b) shows that, in this region, the *kappa* vectors intersect on the ray $\vec{\psi}_\pi$ for all but the fourth *zero* where the intersection is on $\vec{\psi}_0$. Figure 52 helps us appreciate that no matter what intricacies there are in the rotations of $\vec{h}(t)$ and $\vec{\ell}(t)$ there has to be an interval in the rotation of $\vec{\psi}$ between the *zeros*. This is the mechanism underlying "repulsion".

6.6. Measuring the gaps between the zeros

The gaps between the *zeros* can be measured in t, in ψ_0 or in the arguments of $\vec{h}(t)$ and $\vec{\ell}(t)$. We first look at some low *zeros*. Only two vectors are needed for the first 6 *zeros*. In Figure 53, t_3, t_4 and t_5 are seen in the interval $t = 25$ to $t = 33$. The two vectors of $P(h(t))$, $\vec{\mathcal{R}}_1$ and $\vec{\mathcal{R}}_2$ are red arrows and the two vectors of $P(\ell(t))$, $\vec{\mathcal{L}}_1$ and $\vec{\mathcal{L}}_2$ are blue arrows. There is a symmetrical curtate cycloid-like curve in $\ell(t)$ generated when $\vec{\mathcal{L}}_2$ is directed back on $\vec{\mathcal{L}}_1$ between t_4 and t_5.

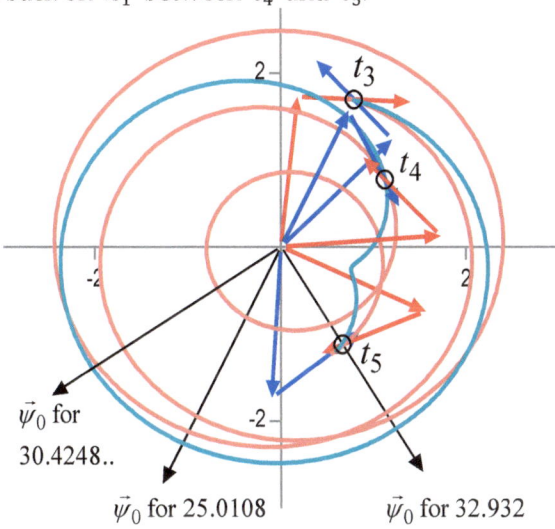

Figure 53. In red $h_{l,\kappa}(t,x)$ and in blue $\ell_{l,\kappa}(t,x)$ from $t = 25$ to $t = 33$ meet for three *zeros*; t_3, t_4 and t_5 with $\kappa = 2$, $l = 3$. Between t_3 and t_4 the ray $\vec{\psi}_0$ completes -2.08 rotations as does $\arg(h(t))$ whilst $\arg(\ell(t))$ completes only -1.08 rotations. Between t_4 and t_5 the ray $\vec{\psi}_0$ completes -0.7604 rotations whilst $\arg(h(t))$ completes -1.2604 rotations and $\arg(\ell(t))$ only -0.2604 rotations i.e. 1/2 rotation more and 1/2 rotation less than $\vec{\psi}_0$.

In Figure 53 t_5 sits on $\vec{\psi}_0$ whilst t_3 and t_4 sit on $\vec{\psi}_\pi$ and consequently the gaps in the arguments of the functions need not equate. Appreciating these intricacies, we now turn to the distributions.

6.7. The distribution of the gaps between neighbouring zeros

The distribution of the gaps between neighbouring *zeros* is compared with a Poisson distribution. The density of the *zeros* rises with t. Later we normalise the gaps between the *zeros* but first we define the gap between neighbouring *zeros* to be

$$\Delta t_i = t_i - t_{i-1} \text{ for } i > 2 \tag{99}$$

with a simple mean gap $\tilde{\Delta t_i}$ over the interval from t_a to t_b with $n = 1 + b - a$ defined as

$$\tilde{\Delta t_i} = \frac{1}{n} \sum_{i=a}^{b} \Delta t_i. \tag{100}$$

The variance[1] σ^2 or second moment of the distribution is

$$\sigma^2 = \frac{1}{n-1} \sum_{i=a}^{b} \left(\Delta t_i - \tilde{\Delta t_i}\right)^2, \tag{101}$$

the third moment, representing skewness, is

$$M_3 = \sum_{i=a}^{b} \frac{\left(\Delta t_i - \tilde{\Delta t_i}\right)^3}{\sigma^3 n} \tag{102}$$

and the fourth representing kurtosis (without the commonly applied subtraction of 3) is

$$M_4 = \sum_{i=a}^{b} \frac{\left(\Delta t_i - \tilde{\Delta t_i}\right)^4}{\sigma^4 n}. \tag{103}$$

We now imagine a short interval in t over which we ignore the increase in the density of *zeros*. If the Poisson distribution held for a set of *zeros* in that interval, there would be independence of the *zeros* with no repulsion. The probability of finding a *zero* in an interval would be proportional to the length of the interval. For Poisson, the probability P of finding k *zeros* in an interval is

$$P(k) = \frac{\lambda^k e^{-\lambda}}{k!} \tag{104}$$

with $k \in \mathbb{N}_0$ (Roman k not Greek *kappa* κ).

Figure 54 illustrates, in grey, a short interval in t of length 5327.66 running from $t_i = 1{,}126{,}903.7347$ to $t_i = 1{,}132{,}231.4002$ including 10,260 *zeros* and embracing 10,259 gaps where $\kappa = 424$. These are single gaps

---

[1] The standard deviation uses σ without inviting confusion with $\text{Re}(s)$.

between immediately neighbouring *zeros*. In this interval, *kappa* was constant at 424. The distribution is leptokurtic with positive skewness characterised with a mean $\tilde{\Delta}t_i = 0.5193$, $\sigma^2 = 0.2127$, $M_3 = 0.4680$ and $M_4 = 3.1268$. Figure 54 has in red the distribution of gaps for the 7221 *zeros* for *kappa* = 314 with a larger average gap of $\tilde{\Delta}t_i = 0.546$.

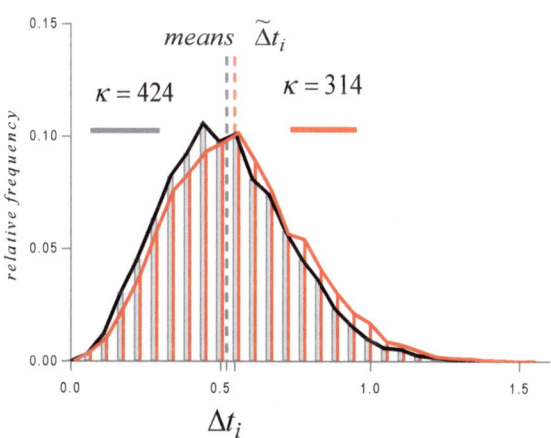

Figure 54. Histograms of the intervals between 10,259 consecutive *zeros* in grey for $\kappa = 424$ and the 7,221 *zeros* for $\kappa = 314$ in red. Note the increased density in *zeros* as t rises, evident with the difference in means and note also the asymmetric distributions.

The interval between *zeros* in the set of 10,259 *zeros* was tested against an expectation of randomness. The number of *zeros* in each of 2,767 successive equal intervals in t of length 1.9256 was determined and plotted in grey in Figure 55 and compared with the expectation of the much broader and flatter Poisson distribution in blue.

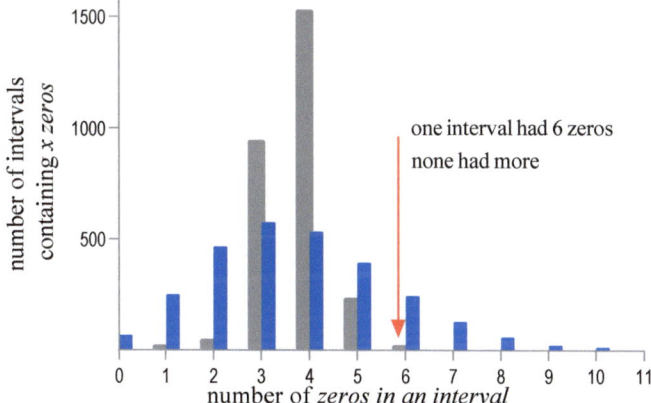

Figure 55. A plot of the number of intervals, of length 1.9256, containing from 0 to 11 *zeros*. In grey the number of intervals containing 2, 3, 4, 5, 6 *zeros* and in blue the expectation from a Poisson distribution. The absent tail of the distribution for the *zeros* on the right-hand side is an important observation; fewer intervals with 5 *zeros*, only one with 6 and none with more.

Figure 55 on page 85 shows that a Poisson model (in blue) does not apply to the distribution of the *zeros* (in grey). The Poisson model with $\lambda = 10259/2767 = 3.707$ and the data appear in Table A8 in Appendix A8 (page 120). The *zeros* are more evenly distributed than a random process would generate. Out of the 2,767 intervals there is only one interval with 6 *zeros* in it and none with more. In contrast a random Poisson process would generate 473 intervals (17.1% of intervals) with 6 or more *zeros* in them. The distribution of gap sizes in Figure 55 (page 85) is also clearly not Gaussian.

The same data are reproduced in grey in Figure 56(a) opposite on page 87 with the average gap scaled up to 1 and with the relative frequency on the right-hand axis. For comparison, the pair-correlation function of relevance to the GUE is in red.

This mean gap (with a delta tilde) $\tilde{\Delta}t_i$ used for the short interval in t in Figure 54 above on page 85, is to be distinguished from the normalised gap, (designated here with a delta hat) $\hat{\Delta}t_{i,a}$, that we can use for gap sizes $a \geq 1$ and which is defined as follows;

$$\hat{\Delta}t_{i,a} = \frac{t_i - t_{i-a}}{2\pi} \ln\left(\frac{t_i}{2\pi}\right). \tag{105}$$

In addition to immediate neighbours, the pair-correlation function considers pairs of *zeros* separated by more than one gap. Consequently, nearly all real gaps are possible and the function has an asymptote of 1. The early damped oscillations of the pair-correlation function have local maxima at multiples of the mean interval for a single gap.

It is easy to understand why the damping increases with larger gap sizes. As a rises the distribution of gap sizes for that a will become less skewed and if a high gap size (e.g. 100) is entertained on its own, then the distribution will be close to Gaussian.

This drift to a Gaussian-like distribution as a rises is shown in Figure 56(b) opposite on page 87. However, no matter how large t becomes we expect the distribution of the single gaps ($a = 1$), to be asymmetric in a similar way to that shown in Figure 56(a).

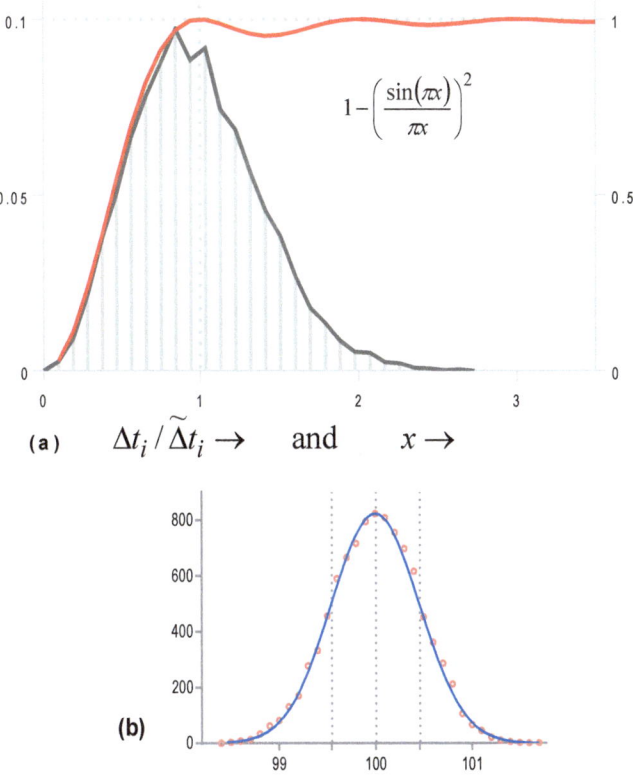

Figure 56. (a) In grey, a histogram of the relative frequency (left-hand axis), against gap size (x-axis), for immediate neighbours with the mean gap scaled to 1. In red the pair-correlation function $1 - \left(\frac{\sin(\pi x)}{\pi x}\right)^2$ using the right-hand axis. **(b)** Red dots show the histogram of the distribution of 9594 normalised gaps $\hat{\Delta}t_{i,a}$ for $a = 100$ in an interval from $t_i = 623{,}025.85$ to $t_i = 1{,}132{,}463.58$. The blue curve is a Gaussian curve with mean 100, and $\sigma^2 = 0.4572$ and an amplitude of 822.

In Figure 56(b) the 9594 normalised gaps have a mean of 100, $\sigma^2 = 0.4572$, skewness 0 and kurtosis -3.1375. The pair-correlation function is the probability of finding another event in a small interval near x; that event being an immediate neighbour or one separated by any number of intervening events. The Gaussian plot in blue is, as expected, a good fit for $a = 100$.

6.8. Summating the histograms to give the pair-correlation curve

Figure 57 shows gaps between *zeros* normalised as in Equation 105 (page 86) with $a \in \{1, 2, 3, 4\}$. The first 2,001,052 *zeros* have 2,001,051 gaps with $a = 1$, 2,001,050 gaps with $a = 2$ etc. The black curve in Figure 57 is the sum of the 4 histograms.

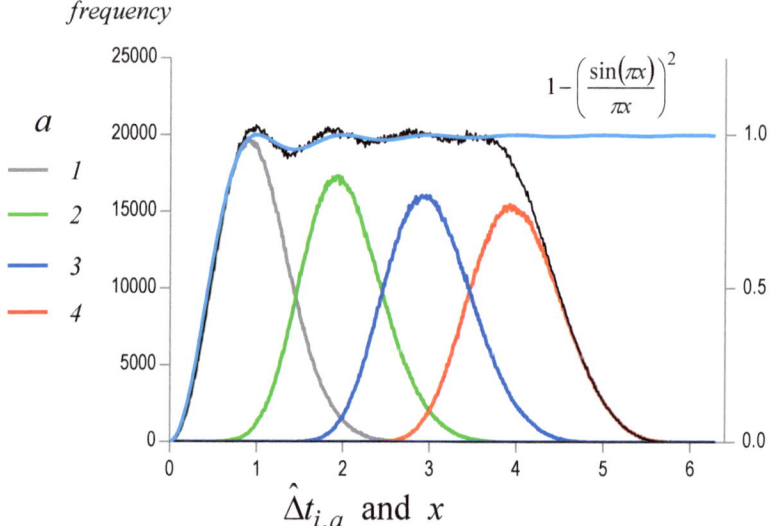

Figure 57. Histograms of the frequencies (left-hand axis) of the normalised gap size $\hat{\Delta}t_{i,a}$, for the paired *zeros*; $a = 1$ gap in grey, $a = 2$ gaps in green, $a = 3$ gaps in blue and $a = 4$ gaps in red, for the first 2,001,052 *zeros*. On the right-hand axis is the pair-correlation function against x plotted in light blue. The black curve is the sum of the four histograms.

When the eigenvalues of the GUE are normalised their pair-correlation function follows the light blue curve of Figure 57 (right-hand axis). The pair-correlation function has local maxima at 1, 2, 3, 4, ... and local minima at 3/2, 5/2, 7/2, ... The peak of the histogram for $\hat{\Delta}t_{i,1}$ lies to the left of 1 and that for $\hat{\Delta}t_{i,2}$ to the left of 2 etc, this is because the third moments of the $\hat{\Delta}t_{i,a}$ distributions for lower values of a are positive. Consequently, the peaks and troughs of the summated histograms precede the local maxima and minima of the pair-correlation function. Even though, as a rises, the distribution tends towards a symmetrical Gaussian distribution the summation is not exactly equivalent to the pair-correlation function.

6.9. A comparison of the normalised interval in t, with the rotation of $\vec{\psi}_0$

We now simplify the notation for the normalised gap between immediately neighbouring *zeros* and use $\hat{\Delta}t_i$, looking backwards in t over one gap and dropping the subscript a so that

$$\hat{\Delta}t_i = \hat{\Delta}t_{i,1} = \frac{t_i - t_{i-1}}{2\pi} \ln\left(\frac{t_i}{2\pi}\right). \qquad (106)$$

This is compared with $\Delta\psi_i/(2\pi)$ being completed rotations of $\vec{\psi}_0$ using

$$\Delta\psi_i = \psi_i - \psi_{i-1} + \pi([v_i/l] - [v_{i-1}/l])$$

$$\text{or } \Delta\psi_i = \psi_i - \psi_{i-1} + \pi\left(\left[\frac{t_i}{2\pi}\right] - \left[\frac{t_{i-1}}{2\pi}\right]\right) \qquad (107)$$

in which $v_i = \left(\left(e^{\frac{2\pi}{t_i l}} - 1\right)^{-1} + \left(1 - e^{-\frac{2\pi}{t_i l}}\right)^{-1}\right)/2$ being simply the v_1 for t_i.
In this way there is an additional 1/2 a rotation every time there is an integral change in $[v_i/l]$ and the function is limited to the set where ψ_i and ψ_{i-1} are determined for the same value of *kappa*. The relationship we expect to find for large t will be

$$\Delta\psi_i/(2\pi) = -\hat{\Delta}t_i \qquad (108)$$

and in the limit, we expect $\Delta\psi_i/(2\pi) \to 1$ as t rises. This is examined in Figure 58(a) with the first 1000 *zeros* and in Figure 58(b) where 10,400 *zeros* at higher values of t are used.

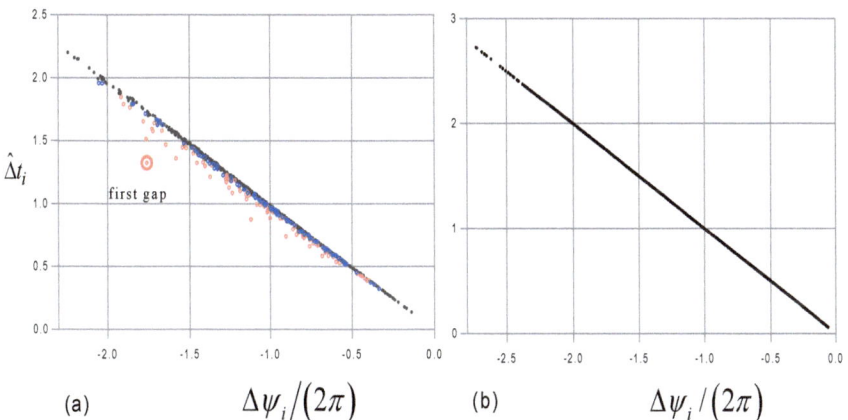

Figure 58. Dot plots of $\hat{\Delta}t_i$ against the number of rotations $\Delta\psi_i/(2\pi)$ of the line of reflection $\vec{\psi}$ between those *zeros*. Being clockwise, $\Delta\psi_i$ is negative. **(a)** Gaps for the first 100 *zeros* are in pink, the second hundred are in blue and then the remaining gaps till t_{1000} are in black. **(b)** Gaps for 10,400 *zeros* from 1,127,063.73... to 1,132,463.581... .

Figure 58 shows a near perfect correlation $\Delta\psi_i/(2\pi) = -\hat{\Delta}t_i$ as t rises, remembering that this acknowledges the step changes when gaps straddle changes in $[v_i/l]$. We next turn to the rotations of $\vec{\ell}(t)$ and $\vec{h}(t)$.

6.10. A comparison of the rotation of $\vec{\psi}_0$ with the rotations of $\vec{\ell}(t)$ and $\vec{h}(t)$ between the zeros

For this section the gaps for 10,400 zeros from 1,127,063.73... to 1,132,463.581..., as used in Figure 58(b), were examined in detail at small intervals between each zero to determine the progress of the arguments of the functions $\arg(\vec{\ell}(t))$ and $\arg(\vec{h}(t))$ between each zero. This was necessary to account for the cycloid-like behaviours described above. Simply knowing the reduced arguments of the functions at the zeros would have been insufficient. The rotations were designated as follows

$$\Delta\arg(\vec{\ell}(t_i)) = \arg(\vec{\ell}(t_i)) - \arg(\vec{\ell}(t_{i-1})) \tag{109}$$

$$\Delta\arg(\vec{h}(t_i)) = \arg(\vec{h}(t_i)) - \arg(\vec{h}(t_{i-1})). \tag{110}$$

The findings are summarised in Figure 59 as rotations by dividing by 2π.

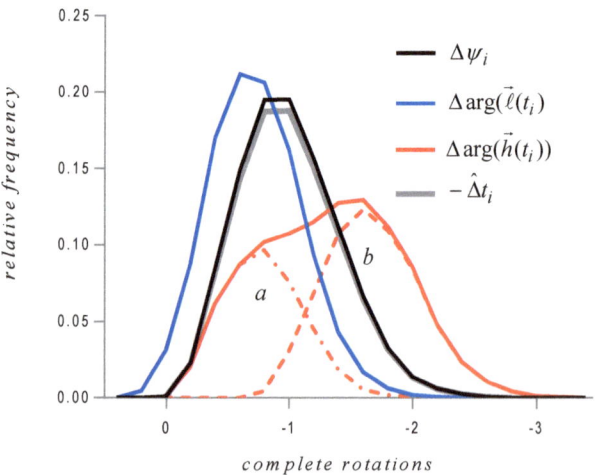

Figure 59. Curves following histograms of relative frequencies against rotations for the gaps between neighbouring zeros. The intervals in the rotation of $\vec{\psi}_0$ between zeros are $\Delta\psi_i$ (black) which follow $-\hat{\Delta}t_i$ (grey). In blue is $\Delta\arg(\vec{\ell}(t_i))$ and in red $\Delta\arg(\vec{h}(t_i,))$. The latter is composed of two main subpopulations (a) in which $\Delta\arg(\vec{\ell}(t_i)) = \Delta\arg(\vec{h}(t_i))$ and (b) in which $\Delta\arg(\vec{\ell}(t_i)) = \Delta\arg(\vec{h}(t_i)) + 2\pi$.

Figure 59 shows a histogram with $\Delta\psi_i$ (in black) following $-\hat{\Delta}t_i$ (grey), as in Figure 58(b). The minor differences relate to the smaller set for $\Delta\psi_i$ which omits gaps at changes in *kappa*. Between paired *zeros* $\vec{\ell}(t)$ and $\vec{h}(t)$ have to rotate through the same argument, when presented in a reduced form $[0, 2\pi)$ since they are together at the beginning and the end of the interval. An example of a gap contributing to Figure 59 (a) in red, being less than one complete clockwise rotation whilst $\Delta\arg(\vec{\ell}(t_i)) > 0$ (anticlockwise), was seen in Figure 47 and Figure 49 (on pages 78 and 80) between t_c and t_d. The vectors complete the same change in unreduced arguments. Sometimes, as in curve Figure 59 (b) in red, $\Delta\arg(\vec{\ell}(t_i)) = \Delta\arg(\vec{h}(t_i)) + 2\pi$ which describes $\vec{h}(t)$ completing an additional ↻ rotation so lapping $\vec{\ell}(t)$ and catching it up once more. An example of this was seen between t_b and t_c in Figures 47 and 49, pages 78 and 80.

7. The *Beta* Function: the Dirichlet *beta* function satisfies the Generalised Riemann Hypothesis

The Dirichlet or Catalan *beta* function

$$\beta(s) = \sum_{n=1}^{\infty} (-1)^{n-1} (2n-1)^{-s} \tag{111}$$

is also recognisable as a Dirichlet L-series defined as $L(s, \chi) = \sum_{n=1}^{\infty} \chi(n) n^{-s}$ with the Dirichlet character χ_1 having modulus 4;

$$L(s, \chi_1) = \beta(s) \text{ modulus 4.} \tag{112}$$

Since all integers differ from a multiple of 4 by either 0, 1, 2 or 3 all the primes > 2 can be placed into two groups; those where $4|(p-1)$ and those where $4|(p-3)$. Then in a similar way to that in which Euler's *zeta* $\zeta_\infty(s) = \sum_{n=1}^{\infty} n^{-s}$ can be shown to equate to Euler's product over the primes $\prod_{p=primes} \frac{1}{1-p^{-s}}$ in encoding the FTA, the Dirichlet *beta* series can be factorized as follows:

$$\beta(s) = 1^{-s} - 3^{-s} + 5^{-s} - 7^{-s} + 9^{-s} \dots$$

$$3^{-s}\beta(s) = 3^{-s} - 9^{-s} + 15^{-s} - 21^{-s} + 27^{-s} \dots$$

$$(1 + 3^{-s})\beta(s) = 1 + 5^{-s} - 7^{-s} - 11^{-s} + 13^{-s} + 17^{-s} \dots$$

$$5^{-s}(1 + 3^{-s})\beta(s) = 5^{-s} + 25^{-s} - 35^{-s} - 55^{-s} + 65^{-s} + 85^{-s} \dots$$

$$(1 - 5^{-s})(1 + 3^{-s})\beta(s) = 1 - 7^{-s} - 11^{-s} + 13^{-s} + 17^{-s} - 19^{-s} - 23^{-s} \dots$$

proceeding until it is clear that

$$\beta(s) = \prod_{p \equiv 1 \bmod 4} \frac{1}{1-p^{-s}} \prod_{p \equiv 3 \bmod 4} \frac{1}{1+p^{-s}}. \tag{113}$$

In this way Dirichlet's *beta* has a relationship with the primes.

The series has a Functional Equation

$$\beta(1-s) = 2^s \pi^{-s} \sin\left(\frac{\pi s}{2}\right) \Gamma(s) \beta(s) \tag{114}$$

and so if there is a zero at a t_u with $\text{Re}(s) = \sigma_\alpha$ then, since we can ignore the sign of t, there is also a zero at $\text{Re}(s) = \sigma_\beta$ for t_u with $0 < \sigma_\alpha < 1/2 < \sigma_\beta$ and $\sigma_\alpha + \sigma_\beta = 1$. A partial series $\beta_n(s)$ has

$$\text{Re } \beta_n(s) = \sum_{m=1}^{n} (-1)^{m-1} (2m-1)^{-\sigma} \cos(t \ln(2m-1))$$

and $\text{Im } \beta_n(s) = -\sum_{m=1}^{n} (-1)^{m-1} (2m-1)^{-\sigma} \sin(t \ln(2m-1)), \tag{115}$

which facilitate the plotting of $P(\beta_n(s))$ in $\mathbb{R}^2$. Some objects used for *zeta*, *eta* and RH are redefined for *beta*, though we do not need l.

7.1. The paired pathways $P(\beta_n(s))$ and $P(h_r(s))$ start from the origin

The pathway $P(\beta_n(s))$ is a plot of the sequential vectors representing the partial series $\beta_n(s)$;

$$\beta_n(s) = \sum_{m=1}^{n} (-1)^{m-1}(2m-1)^{-s} \text{ and } P(\beta_n(s)) \equiv \sum_{m=1}^{n} \vec{m}$$

with the vectors $\vec{m}$ plotted in $\mathbb{R}^2$ having

$$|\vec{m}| = (2m-1)^{-\sigma} \text{ and } \arg(\vec{m}) = -t\ln(2m-1) + \phi$$

with $\phi = 0$ if $2 \nmid m$ and $\phi = \pi$ if $2 | m$. $\tag{116}$

A positive integer ν_β, locates the $\vec{m}$ nearest the *inflection point* in the pathway of a smooth curve that follows the final paired pseudo-spiral $P(\mathcal{R}_1)$ of $P(\beta_n(s))$ with

$$\nu_\beta = \left[\frac{1}{2}\left((e^{\pi/t} - 1)^{-1} + (1 - e^{-\pi/t})^{-1}\right)\right] \tag{117}$$

and that $\vec{m}$ will have a magnitude of $2\nu_\beta - 1$.

A second convergent vector series $h_r(s)$ and its $P(h_r(s))$ summate an ordered set of $\vec{\mathcal{R}}_r$ using the index r to a final term of r. A vector $\vec{\mathcal{R}}_r$ represents the magnitude of the *principal-axis* of a $P(\mathcal{R}_r)$ and is an object in its own right. It has a magnitude $|\vec{\mathcal{R}}_r|$ being

$$|\vec{\mathcal{R}}_r| = (2v_\beta - 1)^{(1/2-\sigma)} r^{(\sigma-1)}, \tag{118}$$

with an argument

$$\arg(\vec{\mathcal{R}}_r) = \theta + t\ln\left(\frac{2r-1}{2v_\beta - 1}\right)$$

with $\theta = \pi/4$ if $2|v_\beta$ and $\theta = 5\pi/4$ if $2 \nmid v_\beta$. (119)

The parameter θ accounts for the alternation of terms. The $\vec{v}_\beta$ faces forwards for odd values and is reversed for even values. The series to r terms, $h_r(s)$ is defined as;

$$h_r(s) = \sum_{r=1}^{r} \vec{\mathcal{R}}_r \text{ with } \vec{\mathcal{R}}_r \text{ following } r \in \mathbb{N}, \tag{120}$$

and the series $h_\kappa(s)$, to *kappa* terms is simply $h_\kappa(s) = \sum_{r=1}^{\kappa} \vec{\mathcal{R}}_r$. We then have an approximation for the *beta* function as the difference between the two finite series to *kappa* terms

$$\beta(s) \approx \beta_\kappa(s) - h_\kappa(s). \tag{121}$$

This difference can be refined by taking account of the intersection of the $\vec{m}_\kappa$ and $\vec{\mathcal{R}}_\kappa$ with a function $x = f(\kappa)$ having $x \in \{x : 0 < x \leq 1\}$ but whose finer details are not important to the proof. A vector function closely related to $\beta(s)$ which we will designate $\beta(s, x)$ is

$$\beta(s, x) = \left(\beta_{\kappa-1}(s) + x\vec{\kappa}_\beta\right) - \left(h_{\kappa-1}(s) + x\vec{\mathcal{R}}_\kappa\right), \tag{122}$$

whose symmetry breaking either side of $\sigma = 1/2$ is easily appreciated from $P(\beta_r(s))$, $P(\beta_\kappa(s))$ and $P(h_\kappa(s))$ which forces $\beta(s)$ to obey the GRH. A line of reflection $\vec{\psi}$, is made of two rays $\vec{\psi}_0$ and $\vec{\psi}_\pi$ of unspecified magnitude and having arguments ψ_0 and ψ_π. The argument ψ_0 is

$$\psi_0 = \arg(\vec{\psi}_0) = \pi/8 - (t/2)\ln(2v_\beta - 1). \tag{123}$$

The principal argument ψ_π is displaced clockwise from ψ_0 and so $\psi_\pi = \psi_0 - \pi$. The pathway $P(\beta_\infty(s))$ has an $\vec{m} = \vec{\tau}$ in the *distal* pseudo-spiral of its $P(\mathcal{R}_1)$ for which $m = \tau$;

$$\tau = \left[\left(e^{\pi/t} - 1\right)^{-1} + \left(1 - e^{-\pi/t}\right)^{-1}\right]. \tag{124}$$

The vector series $P(\beta_n(s))$ indexed by m and $P(h_r(s))$ indexed by r have *proximal* and *distal* parts separated at terms designated *kappa*. A $\vec{m}$ for which $m = \kappa$ is designated $\vec{\kappa}$ and a $\vec{\mathcal{R}}_r$ for which $r = \kappa$ is designated $\vec{\mathcal{R}}_\kappa$. *Kappa* $\in \mathbb{N}_1$ is defined as

$$\kappa = \left[\frac{1}{2}\left((e^{2\pi/t} - 1)^{-1/2} + (1 - e^{-2\pi/t})^{-1/2}\right)\right]. \tag{125}$$

A residual, *kappa dot* written $\dot{\kappa} \in \mathbb{R}$, with $-1/2 \leq \dot{\kappa} \leq 1/2$ is simply,

$$\dot{\kappa} = \frac{1}{2}\left((e^{2\pi/t} - 1)^{-1/2} + (1 - e^{-2\pi/t})^{-1/2}\right) - \kappa. \tag{126}$$

Kappa dot is the domain of a function indicated $x = f(\dot{\kappa})$ with $0 < x \leq 1$ being the intersection of $\vec{\kappa}$ and $\vec{\mathcal{R}}_\kappa$ at the zeros.

A set of zeros of $\beta(s)$ for $t \leq 5250$ at $\sigma = 1/2$ were located by finding minima with very low magnitudes in $\beta_\tau(s)$, or at very low values of t, using $\beta_{2\tau}(s)$; a sample appear in Appendix A9 (page 121–122). Intervals in t of 0.001 were examined from $t = 6$ to $t = 4600$ and then in intervals of 0.0001 from $t = 4600$ to $t = 5250$. Since any region of t can easily be surveyed for *zeros* a few higher *zeros* were also examined. Symmetrical pathways for a *beta* zero are illustrated in Figure 60.

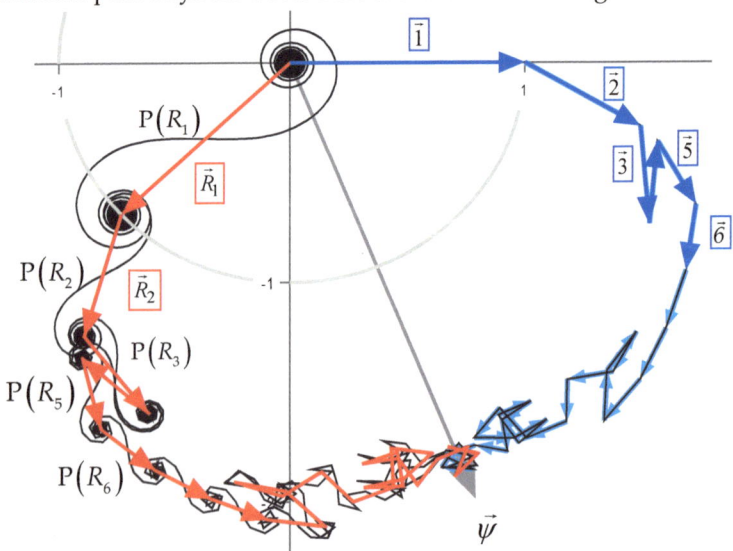

Figure 60. $P(\beta(1/2 + i5247.848260))$ for the 5944th zero of *beta* in black. In blue the $P(\beta_\kappa(s))$ with $\kappa = 29$ with the first 6 vectors emboldened. In red $P(h_\kappa(s))$ retraces the *distal* $P(\beta(s))$. The line of reflection $\vec{\psi}$ is shown and $P(\mathcal{R}_1)$ to $P(\mathcal{R}_3)$ and $P(\mathcal{R}_5)$ and $P(\mathcal{R}_6)$ are labeled. Symmetry breaking forces $\beta(s)$ to obey the GRH.

8. Discussion

8.1. The Riemann Hypothesis

In 1826 Niels Abel said it was a disgrace to base a proof on a divergent series, *"Les séries divergentes sont en général quelque chose de bien fatal et c'est une honte qu'on ose y fonder aucune démonstration."* which is often misquoted as "divergent series being an invention of the devil". In this monograph, Euler's divergent *zeta* series is foreshortened to *tau* terms as $\zeta_\tau(s)$, with *tau* related to t/π. The foreshortening implies a focal point in the final pseudo-spiral of $P(\zeta_\tau(s))$ — with $P(f(s))$ being the operation of plotting the vector series $f(s)$ in $\mathbb{R}^2$.

Euler's *zeta* has a *proximal* pathway to *kappa* terms, with *kappa* related to $\sqrt{t/(2\pi)}$, and a *distal* pathway of paired pseudo-spirals $P(\mathcal{R}_r)$ whose *principal-axes* equate to vectors $\vec{\mathcal{R}}_r$ of a complementary finite series $h_{1,\kappa}(s)$. A symmetry argument requires that for $\zeta_\tau(s) \approx 0$ we have $\text{Re}(s) = 1/2$. A modified Dirichlet *eta* series, designated $\eta_l(s)$, converges through multiplication of every l^{th} term by $(1-l)$ and the nontrivial *zeros* of $\zeta(s)$ are also the *zeros* of $\eta_l(s)$ and of $\eta_{l+1}(s)$. The *proximal* pathway of $\eta_{l,\kappa l}(s)$ is closely followed by the sum of the finite series $\zeta_{\kappa l}(s) + \ell_{l,\kappa}(s)$, and if $l > \kappa$ the distal pathway has a relationship with a finite series $h_{l,\kappa}(s)$.

This monograph has four arguments for the Riemann Hypothesis. The infinite series $\eta_l(s)$ with $l > \kappa$ is closely followed by the sum of three finite series: $\eta_l(s) = \zeta_{\kappa l}(s) + \ell_{l,\kappa}(s) - h_{l,\kappa}(s)$. The first, $\zeta_{\kappa l}(s) \approx 0$ when $s \in \{\rho\}$ if $\kappa l \approx \tau$, allowing $\zeta_{\kappa l}(s)$ to be placed to one side. Focus then turns to $\lambda_l(s) = \ell_{l,\kappa}(s) - h_{l,\kappa}(s)$. For any given $t > 0$, as long as $l > \kappa$, the matched vectors of the two series of $\lambda_l(s)$ share mirror image arguments about a line of reflection $\vec{\psi}$ for all σ, but their magnitudes equate only when $\sigma = 1/2$. These geometric constraints limit the nontrivial zeros to the *critical line* as hypothesised by Riemann — forming the first argument.

The geometries of $P(\ell'_{l,\kappa}(s))$ and $P(h'_{l,\kappa}(s))$ require the zeros of the differential $\eta'_l(s)$ to have $\sigma > 1/2$, a constraint which is shown to preclude any *zero* of *zeta*, hypothesised to lie off the *critical line*, from having a partner the other side of the *critical line*. Since the Functional Equation has to be satisfied this behaviour of the differentials forms the second argument.

If a hypothetical loop in $\eta_l(\sigma)$ for a fixed t were to generate a double-point, and that double-point were to be imagined to be at the origin, then a change in l to $l+1$ would displace the double-point from zero or deny its existence in $\eta_{l+1}(\sigma)$, and since all nontrivial zeros have to exist for all values of l that t would not be the imaginary part of any element of the set $\{\rho\}$—so completing the third argument.

This third argument can be restated as follows: A model for two *zeros*, at the same t_u, exists in $\eta_l(s)$ when $t_i \cong 2\pi k/\ln(l)$ with $k \in \mathbb{N}_1$ since there will be a loop in $\eta_l(\sigma)$ for fixed t_i with a double-point near the origin giving $\eta_l(1/2 + it_i) = 0 \cong \eta_l(1 + it_i)$. However, the mechanics of these two convergence (for the same t) are sensitive to l and are driven by the principal orientation of the pathways in relation to their final collapse to $(1 + 0i)$ as $\text{Re}(s) \to \infty$. Were there a mechanism, capable of generating $\eta_l(\sigma_\alpha + it_i) = 0 = \eta_l(\sigma_\beta + it_i)$, which would disprove the Riemann Hypothesis, it would have to be a mechanism for which this double-point was unaffected by changes in l. No mechanism is capable of generating loops in $\eta_l(\sigma + it_u)$ which are stable under changes in l and which can also generate similar loops in $\eta_l(\sigma)$ that only appear at values of l when $t_i \cong 2\pi k/\ln(l)$. This is a reformulation of the third argument.

The fourth argument is that if there were a $\eta_l(\sigma_\alpha + it_u) = 0 = \eta_l(\sigma_\beta + it_u)$ and we then allowed $l \cong e^{2\pi k/t_u}$ there would be a requirement for two loops in $\eta_l(\sigma)$ for $t = t_u$ with a triple-point, formed by the addition of $0 \cong \eta_l(1 + it_u)$ to the two other zeros. No pathway can behave in such a way to create two loops. Changes to $P(\eta_l(\sigma))$, driven by σ, only affect the *proximo-distal* gradient in the magnitudes of the vectors to create retro-flexion or at best a single loop in the plot of $\eta_l(\sigma)$. Figure 61 has three curves, representing three classes of $\eta_l(\sigma)$; (a) those with no zeros, (b) those with one zero, and (c) those with a double-point at zero.

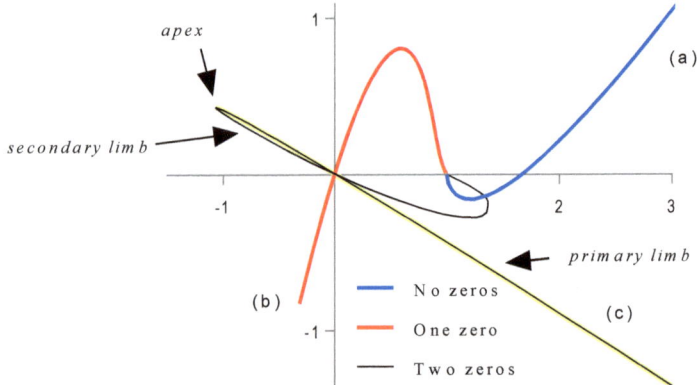

Figure 61. Three curves in $\eta_l(\sigma)$, representing all possible types of curve; all curves head to $(1 + 0i)$ for $\sigma \to \infty$.

A pathway $P(\eta_l(s))$, reaching a point in $\mathbb{R}^2$ on a curve $\eta_l(\sigma)$, undergoing *proximo-distal* shrinkage as σ changes, can only develop at most one loop.

8.2. Remarks on Speiser's corollary

Speiser's corollary has been arrived at in a novel way. Any loop in $\eta_l(\sigma)$ for some t, with a double-point for two real parts $\sigma_a < \sigma_b$, not necessarily at zero, has two limbs. A particle moving in $\mathbb{R}^2$ experiencing σ as time moves swiftly along the primary limb passing the double-point at σ_a before slowing to reach the apex of the loop, it then briefly accelerates, then slows into the secondary limb, and at σ_b it passes through the double-point before heading into the infinite deceleration which leads to $(1 + 0i)$. At the apex, $\frac{\partial \eta_l(\sigma)}{\partial \sigma}$ will be a minimum and the apex will have a value of σ closer to σ_a than σ_b. If t is adjusted in the correct direction the interval between σ_a and σ_b will alter with σ_a rising to $\sigma_{a'}$ and σ_b falling to $\sigma_{b'}$. At some nearby $t = t_c$ the loop vanishes when $\sigma_{a'} = \sigma_{b'} = \sigma_c$ at which point $\frac{\partial \eta_l(\sigma_c)}{\partial \sigma} = 0 = -i\frac{\partial \eta_l(t_c)}{\partial t}$ with $\sigma_b - \sigma_c > \sigma_c - \sigma_a$. If we consider zeros off the *critical line* that refute RH, and satisfy the Functional Equation, we have a hypothetical loop, $\eta_l(\sigma_\alpha + it_i) = 0 = \eta_l(\sigma_\beta + it_i)$, and this loop vanishes at a nearby t_c when $\frac{\partial \eta_l(\sigma_c)}{\partial \sigma} = 0 = -i\frac{\partial \eta_l(t_c)}{\partial t}$ with $\sigma_\beta - \sigma_c > \sigma_c - \sigma_\alpha$ giving $\frac{\partial \eta_l(\sigma_c)}{\partial \sigma} = 0$ only when $\sigma_c < 1/2$. If RH is false the first differential of $\eta_l(s)$ must have at least one zero to the left of the *critical line*. This is a corollary of RH. It is easy to find examples of $\frac{\partial \eta_l(\sigma)}{\partial \sigma} = 0$ when $\sigma > 1/2$ but the geometry of the pathway of the differential necessarily precludes $\frac{\partial \eta_l(s)}{\partial \sigma} = 0$ for $\mathrm{Re}(s) < 1/2$.

The *lambda* function $\lambda_l(s, x)$ has bilateral symmetry, and this limits its own *zeros* to the *critical line* and those *zeros* each have a close relationship to a member of the set $\{\rho\}$. *Lambda* gives some small uncertainty in the $\mathrm{Im}(\rho)$, with each zero hosted in a narrow imaginary interval with the uncertainty related to $x = f(\dot{\kappa})$, but the $\mathrm{Re}(\rho)$ is necessarily limited to $1/2$. There is a relationship between x and $\dot{\kappa}$ that is evident from calculations at low values of t. That relationship has not been fully characterised and has been set aside as it is of no material consequence to the arguments of this monograph, see Appendix A5 page 111.

When a partial Euler's *zeta* function is added to $\lambda_l(s)$ it breaks the symmetry since $\eta_{l,\kappa l}(s) = \zeta_{\kappa l}(s) + \ell_{l,\kappa}(s)$, and it is this symmetry breaking which permits the generation of loops in $\eta_l(\sigma)$ for fixed t. A loop in $\eta_l(\sigma)$ for fixed t has the double-point $\eta_l(\sigma_a + it) = \eta_l(\sigma_b + it)$ and raises the possibility of hypothetical zeros and the existence of a pair of ρ_u. It is this same asymmetry which allows pathway collapse, when σ rises outside the *critical strip*, to drive all pathways towards their final demise at $(1,0)$. Any such mechanism which generates loops in $\eta_l(\sigma)$ for fixed t would be highly sensitive to $\arg(\vec{\mathcal{R}}_1)$ and ψ_0 and so to changes in l. The clarity delivered by the function $\lambda_l(s)$ is in allowing us to realise that it is the symmetry breaking of $\zeta_{\kappa l}(s)$ that permits $\eta_l(1/2 + it_i) = \eta_l(1 + it_i) = 0$ when $t_i = 2\pi k/\ln(l)$, with $k \in \mathbb{N}_1$ (Roman k not Greek κ).

Once $l > \kappa$ the symmetries in $\eta_l(s)$ are evident, and when $l = [\tau/\kappa]$ the term $\zeta_{\kappa l}(\rho) = 0$ allows $\lambda_l(s) = 0$ to reliably specify short regions of the imaginary axis which can host the nontrivial zeros of Riemann's *zeta* function. Importantly, $\lambda_l(s) = 0$ confines the hosting regions to $\operatorname{Re}(s) = 1/2$.

8.3. Remarks on the Riemann-Siegel formula

The Riemann-Siegel Formula (RSF) was discovered by Carl Siegel in 1932 in some of Riemann's manuscripts [14] and it amounts to an approximate Functional Equation which contains two finite Dirichlet series and an error term. The RSF is reproduced in Equations 127 and 128 with $(1/\pi)^{(1/2-s)}\Gamma\left(\frac{1}{2}-\frac{s}{2}\right)/\Gamma\left(\frac{s}{2}\right)$ in place of what might be the more familiar $\gamma(1-s)$ where $\gamma(s) = (\pi)^{(1/2-s)}\Gamma\left(\frac{s}{2}\right)/\Gamma\left(\frac{1}{2}-\frac{s}{2}\right)$,

$$\zeta(s) = \sum_{n=1}^{N} n^{-s} + \frac{\Gamma\left(\frac{1}{2}-\frac{s}{2}\right)}{\Gamma\left(\frac{s}{2}\right)}\left(\frac{1}{\pi}\right)^{(1/2-s)} \sum_{n=1}^{M} n^{s-1} + R(s), \quad (127)$$

$$R(s) = -\frac{\Gamma(1-s)}{2\pi i}\int \frac{(-x)^{(s-1)} e^{-mx}}{e^x - 1} dx. \quad (128)$$

Importantly M and N are typically set at $\left\lfloor\sqrt{t/(2\pi)}\right\rfloor$ which is very close to *kappa* and so the resemblance of the RSF to

$$\eta_l(s) = \ell_{l,\kappa}(s) - h_{l,\kappa}(s) + \zeta_{kl}(s) \quad (129)$$

is of interest when Equation 129 is reformatted as

$$\eta_l(s) \cong -l^{(1-s)}\sum_{n=1}^{n=\kappa} n^{-s} - \sqrt{l}\left(\frac{tl}{2\pi}\right)^{\left(\frac{1}{2}-s\right)} e^{i\frac{\pi}{4}} \sum_{r=1}^{r=\kappa} r^{(s-1)} + \zeta_{kl}(s). \quad (130)$$

8.4. The analogy with the GUE: the pair-correlation function and the repulsion of the zeros

The intervals between the *zeros* of the Riemann *zeta* function have a statistical distribution similar to the gaps between the eigenvalues of an ensemble of random $n \times n$ Hermitian matrices. The processes generating the *zeros* are not random, the *zeros* are not probabilistically independent of one another but repel. That repulsion is explained by the mechanisms described in this monograph.

The two pathways, $P(h_{l,\kappa}(s))$ and $P(\ell_{l,\kappa}(s))$, start from the origin and set off in different directions. If their terminal vectors $\vec{\mathcal{R}}_\kappa$ and $\vec{\mathcal{L}}_\kappa$ intersect over an interval in t there can be a *zero* of $\eta_l(s)$ in that interval. Either $h_{l,\kappa}(s)$ or $\ell_{l,\kappa}(s)$ alone can be used to identify those intervals, and every element of the set $\{t_i\}$ for $\sigma = 1/2$ lies in such an interval. The first vectors of each series rotate clockwise about the origin as mirror images about the line of reflection $\vec{\psi}$. The vector $\vec{h}_{l,\kappa}(s)$ generally rotates clockwise, about the origin, faster than $\vec{\ell}_{l,\kappa}(s)$ and a significant rotation is required before the *kappa* vectors can intersect again after generating a *zero*. This mechanism underlies the repulsion of the *zeros* and provides a model for the pair-correlation function of the GUE.

Appendix

A1. Near-regular polygons and star-polygons

To describe regular star-polygons Ludwig Schläfli's symbol $\{p/q\}$, is traditionally used where p and q are relatively prime [15]; for star-polygons based on the pentagon we have $\{5/2\}$ and $\{5/3\}$. We will hijack this nomenclature and use $\{l/\left(\frac{q}{2}(\bmod\, l)\right)\}$ to describe a family of structures. When $1 = \frac{q}{2}(\bmod\, l)$ we have clockwise near-regular polygons and anticlockwise near-regular polygons when $l - 1 = \frac{q}{2}(\bmod\, l)$. When $l|q$ we have a structure in a pseudo-convergence and if we allow $q = 0$ we have the convergence, otherwise we have near-regular star-polygons.

The pathways between any two pseudo-convergences contain the *inflection point* and exhibit shapes whose geometries can be described as incomplete *near-regular polygons* or *near-regular star-polygons*. Imagine an l-sided regular polygon with its vertices enumerated $\varrho \in \mathbb{N}_0$ and counted clockwise allowing $\varrho > l - 1$, such that a modular arithmetic applies: $\varrho = kl \equiv 0$ with k an integer. A family of new geometries arise when one vertex, at $\varrho = 0$ of the regular l-sided polygon, is connected to another vertex ϱ of the polygon counted in the clockwise direction and the process repeated until the first vertex is reached and the polygon itself or a star-polygon is created. We will allow $\varrho \in \mathbb{N}_0$ and these geometries can be designated $\{l/\varrho\}$ such that a polygon is described when $\varrho = 1$ or $\varrho = l - 1$. If $\varrho = 0$ or $\varrho|l$ we have a special case where the polygon collapses and our structure lies in the pseudo-convergence. Otherwise we have a near-regular star-polygon which is a self-intersecting, near-equilateral near-equiangular structure. For instance, in a regular pentagon, a five-pointed star can be obtained by drawing a line from the first to the third vertex, from the third vertex to the fifth vertex, from the fifth vertex to the second vertex, from the second vertex to the fourth vertex, and from the fourth vertex to the first vertex. Our structures have a missing side which is nearly collinear with the larger neighbouring vectors where $l|m$.

This section illustrates the structures that lie between the lth vectors near the *inflection point* of a series of $P(\mathcal{R}_r)$. The notation is $\{l/\varrho\}$ with l being the number of sides in a regular polygon or the number of vertices in a regular star-polygon, the ϱ is the neighbouring vertex in a clockwise direction to which the next vector is directed.

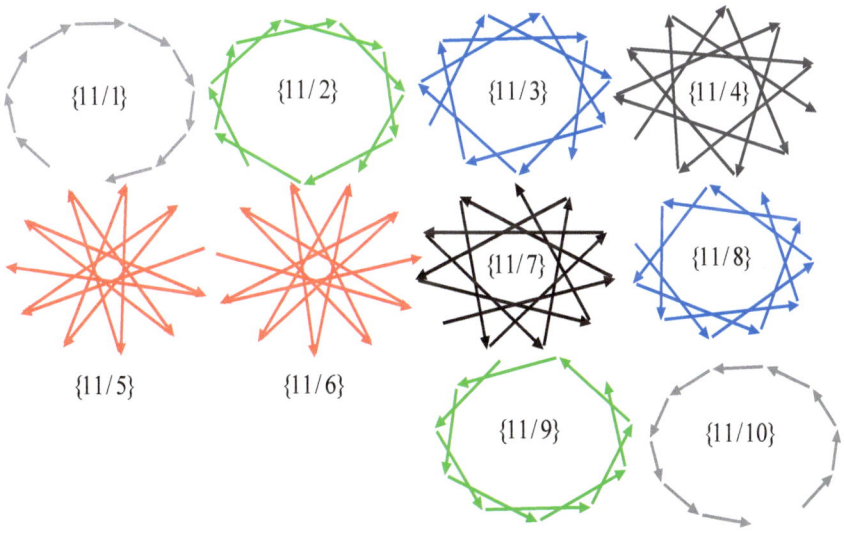

Figure A1. Two near-regular polygons {11/1} and {11/10}, and eight near-regular star-polygons for $l = 11$ are shown. These are taken from P($\eta_l(s)$) near *inflection points* for P($\mathcal{R}_1$) to P($\mathcal{R}_{10}$) for $t = 39096.46$. Structures are not to the same scale.

In Figure A1 the larger l^{th} vectors (where $11|m$), which neighbour the structures, are not shown but would be roughly collinear with each other and in-line with the missing segment of each near-regular structure. The value of $q_{\bar{r}}$ is such that the change in argument at each vector is $q_{\bar{r}}\pi/l$ so that from an lth vector to the next l^{th} vector we have l roughly equal arguments that when summated bring us full circle to give the collinearity seen at inflection. The $\{l/0\}$ structures are collapsed sections that essentially oscillate along one line making no advance in the Argand plane. These structures are located at pseudo-convergences along the pathways. Details appear in Table A1.

Table A1. Polygons and star-polygons.

r	$q_{\bar{r}}$	$\left\{ l / \left(\dfrac{q}{2} \pmod{l} \right) \right\}$	Structural position
		{11/0}	P-C
1	2	{11/1}	Inf
2	4	{11/2}	Inf
3	6	{11/3}	Inf
4	8	{11/4}	Inf
5	10	{11/5}	Inf
6	12	{11/6}	Inf
7	14	{11/7}	Inf
8	16	{11/8}	Inf
9	18	{11/9}	Inf
10	20	{11/10}	Inf
		{11/11} ≡ {11/0}	P-C
11	24	{11/12} ≡ {11/1}	Inf
12	26	{11/13} ≡ {11/2}	Inf

"Inf" means an *inflection point* and "P-C" means a pseudo-convergence

A2. A Note on the collapse of structures either side of $2l|q$

Inherent in the specification of R_l is the absence of the odd values of q either side of the q where $2l|q$. This is illustrated in Figure A2.

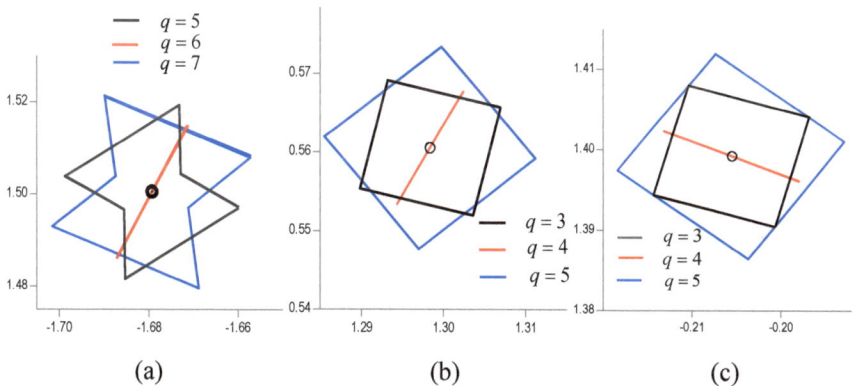

Figure A2. (a) With $l = 3$ and $t = 27{,}016$ the average of neighbouring vectors where $l|m$ is shown with all three averages being in the centre of the black circle. All three values of q have an equivalence when $2l|q$. **(b)** With $l = 2$ the pseudo-convergence is evident for $t = 27{,}016$. **(c)** As in (b) but with $t = 27{,}017$ to show that there has been movement in the Argand plane and rotation of structures but still a shared pseudo-convergence.

A3. Calculations of $|\Delta\hat{\eta}_{q_r}|/|\Delta\hat{\eta}_{q_1}|$ and $(q_{\bar{r}}/2)^{(\sigma-1)}$

Equation 24 (page 19) was compared with the lengths of *principal-axes* determined from averaging as in Equation 13 (page 14). The magnitude of the *principal-axis* being estimated as;

$$\hat{\eta}_q(s) = \left(\eta_{l,a-1}(s) + \eta_{l,a}(s) + \eta_{l,b-1}(s) + \eta_{l,b}(s)\right)/4$$

with $a = l\lfloor m(q)/l \rfloor$ and $b = l(\lfloor m(q)/l \rfloor + 1)$ and $q \in R_l$,

$$\text{giving } |\Delta\hat{\eta}_{q_r}| = |\hat{\eta}_{q_r}(s) - \hat{\eta}_{q_{r+1}}(s)|.$$

In Figure A3 part of a pathway with $l = 2$ is shown, and in Table A2 the ratios of interest.

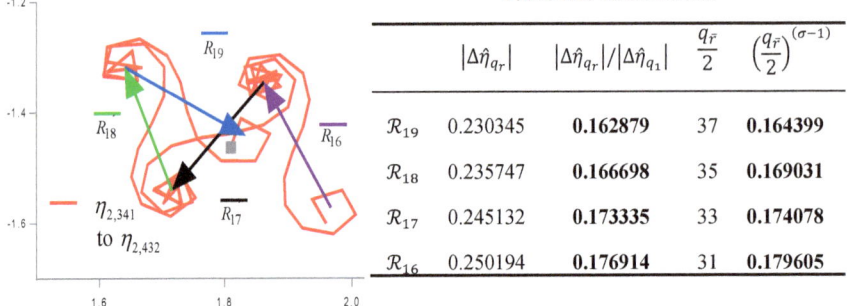

Table A2. Calculations.

	$|\Delta\hat{\eta}_{q_r}|$	$|\Delta\hat{\eta}_{q_r}|/|\Delta\hat{\eta}_{q_1}|$	$\frac{q_{\vec{r}}}{2}$	$\left(\frac{q_{\vec{r}}}{2}\right)^{(\sigma-1)}$
$\mathcal{R}_{19}$	0.230345	0.162879	37	0.164399
$\mathcal{R}_{18}$	0.235747	0.166698	35	0.169031
$\mathcal{R}_{17}$	0.245132	0.173335	33	0.174078
$\mathcal{R}_{16}$	0.250194	0.176914	31	0.179605

Figure A3. A region of $P(\eta_{2,n}(s))$ for $\sigma = 1/2$ and $t_{50,000} = 40433.68739$ running from $m = 341$ (grey square) to $m = 432$ showing the vectors $\vec{\mathcal{R}}_{19}$ to $\vec{\mathcal{R}}_{16}$. The table compares the magnitudes of *principal-axes* by Equation 13 with computations using $(q_{\vec{r}}/2)^{(\sigma-1)}$, see Equation 24 on page 19.

Figure A3 shows part of a pathway to convergence. It illustrates four relatively early vectors in a *distal* pathway where inadequacies of approximations should be most evident; reassuringly the errors are small. If we look closer to convergence we find much smaller errors, see Table A3.

Table A3. Calculations.

	$|\Delta\hat{\eta}_{q_r}|$	$|\Delta\hat{\eta}_{q_r}|/|\Delta\hat{\eta}_{q_1}|$	$q_{\vec{r}}/2$	$(q_{\vec{r}}/2)^{-1/2}$
$\mathcal{R}_5$	0.471419	0.333344	9	0.333333
$\mathcal{R}_4$	0.534537	0.377975	7	0.377964
$\mathcal{R}_3$	0.632446	0.447207	5	0.447214
$\mathcal{R}_2$	0.816487	0.577344	3	0.57735
$\mathcal{R}_1$	1.414213	1	1	1

Table A3 is a comparison of the ratio of averaged summations $|\Delta\hat{\eta}_{q_r}|/|\Delta\hat{\eta}_{q_1}|$ and $(q_{\vec{r}}/2)^{-1/2}$ for the *distal* pathway for $\eta_{l,n}(s)$ for $\sigma = \frac{1}{2}$ for $t_{50,000} = 40433.68739$ and $l = 2$. The summations yield $|\Delta\hat{\eta}_{q_1}|^2 = 1.999998$ consistent with the magnitude of the complimentary vector $|\vec{\mathcal{R}}_1| = \sqrt{2}$, when $l = 2$.

Table A4 below is a comparison for $l = 5$, principally to illustrate the more intricate values of $q_{\vec{r}}/2$ and once again we see $|\Delta\hat{\eta}_{q_1}|^2 = 4.999978$, consistent with the magnitude of the complimentary vector $|\vec{\mathcal{R}}_1| = \sqrt{5}$, when $l = 5$.

Table A4. Calculations.

	$|\Delta\hat{\eta}_{q_r}|$	$|\Delta\hat{\eta}_{q_r}|/|\Delta\hat{\eta}_{q_1}|$	$q_{\tilde{r}}/2$	$(q_{\tilde{r}}/2)^{-1/2}$
$\mathcal{R}_{15}$	0.533496	0.238587	18	0.23570226
$\mathcal{R}_{14}$	0.547925	0.24504	17	0.242535625
$\mathcal{R}_{13}$	0.557137	0.24916	16	0.25
$\mathcal{R}_{12}$	0.596435	0.266734	14	0.267261242
$\mathcal{R}_{11}$	0.6225107	0.278396	13	0.277350098
$\mathcal{R}_{10}$	0.6463426	0.289054	12	0.288675135
$\mathcal{R}_{9}$	0.6736206	0.301253	11	0.301511345
$\mathcal{R}_{8}$	0.7454706	0.333385	9	0.333333333
$\mathcal{R}_{7}$	0.7908195	0.353666	8	0.353553391
$\mathcal{R}_{6}$	0.844766	0.377792	7	0.377964473
$\mathcal{R}_{5}$	0.9128243	0.408228	6	0.40824829
$\mathcal{R}_{4}$	1.1180591	0.500012	4	0.5
$\mathcal{R}_{3}$	1.2910683	0.577385	3	0.577350269
$\mathcal{R}_{2}$	1.5811098	0.707096	2	0.707106781
$\mathcal{R}_{1}$	2.2360624	1	1	1

Table A4 is a comparison of the ratio of averaged summations $|\Delta\hat{\eta}_{q_r}|/|\Delta\hat{\eta}_{q_1}|$ and $(q_{\tilde{r}}/2)^{-1/2}$ for the *distal* pathway for $\eta_{l,n}(s)$ for $\sigma = 1/2$ for $t_{50,000} = 40433.68739$ and $l = 5$. Note that the $q_{\tilde{r}}/2$ column has no integers which have l as a factor.

A4. A comparison of the principal-axes of P($\mathcal{R}_r$) with $|\vec{\mathcal{R}}_r|$

A comparison was made between $|\vec{\mathcal{R}}_r|$ for P($h_l(s)$) at low r and the magnitude of the *principal-axis* of the related P($\mathcal{R}_r$) in $\eta_l(s)$, here designated $|\Delta\hat{\eta}_{q_r}|$. The chosen parameters were $l = 5$ and $t = 40433.69$ as in Figure A3 but with $l = 5$ over a range in σ. The $|\vec{\mathcal{R}}_r|$ was determined by Equation 22 (page 19) $|\vec{\mathcal{R}}_r| = \sqrt{l}\, v^{(1/2-\sigma)}(q_{\tilde{r}}/2)^{(\sigma-1)}$ with $q_{\tilde{r}} \in \chi_l$. The magnitude of the *principal-axis* of the P($\mathcal{R}_r$) was determined from the ends of the axis after summation of the partial series. Averaging was applied as in Equation 13 (page 14). The results appear in Table A5 with σ rising from 0 to 1.1.

Table A5.

r	$q_{\bar{r}}$	Eq 0	$|\Delta\hat{\eta}_{q_r}|$ 0	Eq 0.1	$|\Delta\hat{\eta}_{q_r}|$ 0.1	Eq 0.2	$|\Delta\hat{\eta}_{q_r}|$ 0.2	Eq 0.3	$|\Delta\hat{\eta}_{q_r}|$ 0.3
1	2	401.09604	401.09838	142.06781	142.06841	50.320277	50.3204	17.823392	17.82341
2	4	200.54802	200.54618	76.132256	76.131421	28.90141	28.90104	10.971585	10.97142
3	6	133.69868	133.70715	52.85506	52.858343	20.895175	20.89645	8.2604836	8.260976
4	8	100.27401	100.27624	40.798266	40.799172	16.599501	16.59987	6.7538025	6.753953
5	12	66.049339	66.846111	28.324325	28.322943	12.001127	12.00053	5.0849241	5.08467
6	14	57.299434	57.274438	24.655139	24.644271	10.608759	10.60403	4.5647997	4.562745
7	16	50.137005	50.153472	21.863249	21.870377	9.5339097	9.536994	4.1574531	4.158788
8	18	44.566226	44.573338	19.664252	19.667366	8.6765888	8.677953	3.828429	3.829026
9	22	36.463276	36.432693	16.415053	16.401219	7.3897352	7.383478	3.326714	3.323883
10	24	33.42467	33.467856	15.17863	15.198294	6.8928373	6.901791	3.1301379	3.134215
11	26	30.853541	30.968401	14.123641	14.176347	6.4652949	6.489481	2.9595794	2.970678
12	28	28.649717	28.590153	13.212362	13.185171	6.0931321	6.08072	2.8099638	2.804299
13	32	25.068502	24.984676	11.716225	11.677007	5.4757932	5.457444	2.5592126	2.550627
14	34	23.593884	23.837881	11.09409	11.208776	5.2165561	5.270463	2.4528789	2.478217
15	36	22.283113	22.554743	10.537812	10.666366	4.9833916	5.044233	2.3566745	2.385469
16	38	21.110318	20.918401	10.037313	9.9463647	4.7724363	4.729337	2.269148	2.248724
17	42	19.099811	19.337312	9.1727241	9.2870576	4.4052198	4.46026	2.115616	2.142113
18	44	18.231638	18.345793	8.7966089	8.8515107	4.2442883	4.270693	2.0478327	2.060532

Table A5 (continued).

r	$q_{\bar{r}}$	Eq 0.4	$\|\Delta\hat{\eta}_{q_r}\|$ 0.4	Eq 0.5	$\|\Delta\hat{\eta}_{q_r}\|$ 0.5	Eq 0.6	$\|\Delta\hat{\eta}_{q_r}\|$ 0.6	Eq 0.7	$\|\Delta\hat{\eta}_{q_r}\|$ 0.7
1	2	6.3130274	6.313022	2.236068	2.236062	0.792013	0.79201	0.2805302	0.280529
2	4	4.1650448	4.164976	1.5811388	1.58111	0.6002336	0.600221	0.2278613	0.227856
3	6	3.2656146	3.265805	1.2909944	1.291068	0.5103685	0.510397	0.2017638	0.201775
4	8	2.7479048	2.747966	1.118034	1.118059	0.454892	0.454902	0.1850809	0.185085
5	12	2.1545021	2.154393	0.9128709	0.912824	0.386787	0.386767	0.1638831	0.163875
6	14	1.964169	1.963276	0.8451543	0.844766	0.363658	0.363489	0.1564769	0.156404
7	16	1.8129411	1.813519	0.7905694	0.790819	0.3447437	0.344852	0.1503324	0.150379
8	18	1.6892432	1.689505	0.745356	0.745471	0.3288784	0.328929	0.1451132	0.145135
9	22	1.4976215	1.496341	0.6741999	0.673621	0.3035116	0.303249	0.136635	0.136516
10	24	1.4214413	1.423298	0.6454972	0.646343	0.2931297	0.293515	0.1331145	0.13329
11	26	1.354789	1.359882	0.6201737	0.622511	0.2838932	0.284966	0.1299561	0.130448
12	28	1.2958683	1.293283	0.5976143	0.596435	0.2756012	0.275063	0.1270987	0.126853
13	32	1.196095	1.192078	0.559017	0.557137	0.2612669	0.260387	0.1221079	0.121696
14	34	1.1533691	1.165279	0.5423261	0.547925	0.2550074	0.257639	0.1199071	0.121144
15	36	1.1144849	1.128113	0.5270463	0.533496	0.2492432	0.252296	0.1178685	0.119313
16	38	1.0789107	1.069232	0.5129892	0.508403	0.2439107	0.241737	0.1159721	0.114942
17	42	1.016029	1.028785	0.48795	0.494091	0.234339	0.237295	0.1125418	0.113965
18	44	0.9880617	0.994169	0.4767313	0.479669	0.2300188	0.231431	0.1109821	0.111661

Table A5 (continued).

r	$q_{\bar{r}}$	Eq 0.8	$|\Delta\hat{\eta}_{q_r}|$ 0.8	Eq 0.9	$|\Delta\hat{\eta}_{q_r}|$ 0.9	Eq 1	$|\Delta\hat{\eta}_{q_r}|$ 1	Eq 1.1	$|\Delta\hat{\eta}_{q_r}|$ 1.1
1	2	0.0993635	0.099363	0.0351945	0.035194	0.0124658	0.012466	0.0044154	0.004415
2	4	0.086501	0.086499	0.0328376	0.032837	0.0124658	0.012465	0.0047323	0.004732
3	6	0.0797632	0.079768	0.0315328	0.031534	0.0124658	0.012466	0.0049281	0.004928
4	8	0.0753035	0.075305	0.0306386	0.030639	0.0124658	0.012466	0.0050719	0.005072
5	12	0.0694379	0.069434	0.0294211	0.02942	0.0124658	0.012465	0.0052818	0.005282
6	14	0.0673298	0.067298	0.0289711	0.028957	0.0124658	0.01246	0.0053639	0.005361
7	16	0.0655555	0.065576	0.0285868	0.028596	0.0124658	0.01247	0.005436	0.005438
8	18	0.0640293	0.064039	0.0282521	0.028256	0.0124658	0.012468	0.0055004	0.005501
9	22	0.0615104	0.061457	0.0276908	0.027666	0.0124658	0.012455	0.0056119	0.005607
10	24	0.0604492	0.060529	0.0274509	0.027487	0.0124658	0.012482	0.0056609	0.005668
11	26	0.0594892	0.059715	0.027232	0.027336	0.0124658	0.012513	0.0057064	0.005728
12	28	0.058614	0.058502	0.027031	0.02698	0.0124658	0.012443	0.0057489	0.005738
13	32	0.0570694	0.056877	0.0266724	0.026582	0.0124658	0.012424	0.0058261	0.005806
14	34	0.0563816	0.056963	0.0265112	0.026785	0.0124658	0.012594	0.0058616	0.005922
15	36	0.0557407	0.056425	0.0263601	0.026684	0.0124658	0.012619	0.0058952	0.005968
16	38	0.0551412	0.054653	0.026218	0.025987	0.0124658	0.012356	0.0059271	0.005875
17	42	0.0540484	0.054734	0.0259569	0.026287	0.0124658	0.012625	0.0059867	0.006063
18	44	0.0535479	0.053875	0.0258364	0.025994	0.0124658	0.012541	0.0060147	0.006051

The exercise was repeated for the differentials. A comparison being made between the $\left|\frac{\partial \vec{\mathcal{R}}_r}{\partial \sigma}\right|$ for low values of r in $P(\bar{h}'_l(s))$ and the magnitude of the *principal-axis* of the related $P(\mathcal{R}'_r)$, in $\eta'_l(s)$, for $l = 5$ and $t = 40433.69$ over the same range of real domains using $\left|\frac{\partial \vec{\mathcal{R}}_r}{\partial \sigma}\right| = \sqrt{l}\, v_1^{(\frac{1}{2}-\sigma)} \left(\frac{q_{\bar{r}}}{2}\right)^{(\sigma-1)} \ln\left(\frac{2v_1}{q_{\bar{r}}}\right)$ with $q_{\bar{r}} \in \chi_l$. The magnitude of the *principal-axis* of a $P(\mathcal{R}'_r)$ in $P(\eta'_l(s))$ was determined from the location of the start and finish of the axis after summation of the partial series. Averaging was applied as in a modification of Equation 13, as reproduced here. The magnitude of the *principal-axis* by summation is simply;

$$\widehat{\eta}'_q(s) = \left(\eta'_{l,a-1}(s) + \eta'_{l,a}(s) + \eta'_{l,b-1}(s) + \eta'_{l,b}(s)\right)/4$$

with $a = l\lfloor m(q)/l \rfloor$ and $b = l(\lfloor m(q)/l \rfloor + 1)$ and $q \in R_l$

$$\left|\Delta\widehat{\eta}'_{q_r}\right| = \left|\widehat{\eta}'_{q_r}(s) - \widehat{\eta}'_{q_{r+1}}(s)\right|.$$

The results appear in Table A6 with σ rising from 0 to 1.1.

Table A6.

r	$q_{\bar{r}}$	Eq 0	$\left|\Delta\widehat{\eta}'_{q_r}\right|$ 0	Eq 0.1	$\left|\Delta\widehat{\eta}'_{q_r}\right|$ 0.1	Eq 0.2	$\left|\Delta\widehat{\eta}'_{q_r}\right|$ 0.2	Eq 0.3	$\left|\Delta\widehat{\eta}'_{q_r}\right|$ 0.3
1	2	4162.9998	4162.992	1474.5281	1474.525	522.2756	522.2745	184.9892	184.9888
2	4	1942.4896	1942.457	737.4088	737.3964	279.9355	279.9307	106.2693	106.2674
3	6	1240.7825	1240.851	490.5174	490.5447	193.9158	193.9266	76.6606	76.66486
4	8	901.7396	901.7511	366.8882	366.8935	149.2748	149.2771	60.7350	60.73606
5	12	574.0544	574.0215	243.2287	243.215	103.0568	103.0511	43.6655	43.66313
6	14	483.2138	483.001	207.9198	207.8276	89.4648	89.42488	38.4954	38.47809
7	16	416.1172	416.2511	181.4560	181.5142	79.1274	79.15273	34.5050	34.51606
8	18	364.6328	364.688	160.8891	160.9135	70.9901	71.00085	31.3234	31.32814
9	22	291.0187	290.7733	131.0107	130.8999	58.9784	58.92834	26.5509	26.52828

Table A6 (continued).

r	$q_{\bar{r}}$	Eq 0.4	$\|\Delta\widehat{\eta}'_{q_r}\|$ 0.4	Eq 0.5	$\|\Delta\widehat{\eta}'_{q_r}\|$ 0.5	Eq 0.6	$\|\Delta\widehat{\eta}'_{q_r}\|$ 0.6	Eq 0.7	$\|\Delta\widehat{\eta}'_{q_r}\|$ 0.7
1	2	65.5229	65.52274	23.2081	23.20805	8.220274	8.220254	2.911608	2.911601
2	4	40.3420	40.34127	15.3146	15.31437	5.813753	5.813647	2.207020	2.206979
3	6	30.3061	30.30786	11.9809	11.98158	4.736399	4.73667	1.872436	1.872544
4	8	24.7111	24.71155	10.0541	10.05433	4.090694	4.090785	1.664369	1.664409
5	12	18.5012	18.50023	7.8390	7.838614	3.321420	3.321249	1.407296	1.407225
6	14	16.5640	16.5565	7.1272	7.123997	3.066748	3.065341	1.319576	1.318967
7	16	15.0466	15.05138	6.5614	6.563443	2.861209	2.862114	1.247686	1.24808
8	18	13.8210	13.82311	6.0983	6.099254	2.690798	2.691212	1.187277	1.18746
9	22	11.9527	11.94246	5.3808	5.37624	2.422348	2.420268	1.090492	1.089553

r	$q_{\bar{r}}$	Eq 0.8	$\|\Delta\widehat{\eta}'_{q_r}\|$ 0.8	Eq 0.9	$\|\Delta\widehat{\eta}'_{q_r}\|$ 0.9	Eq 1	$\|\Delta\widehat{\eta}'_{q_r}\|$ 1	Eq 1.1	$\|\Delta\widehat{\eta}'_{q_r}\|$ 1.1
1	2	1.031287	1.031284	0.365280	0.365279	0.129382	0.129381	0.045827	0.045827
2	4	0.837830	0.837814	0.318057	0.318051	0.120741	0.120739	0.045836	0.045835
3	6	0.740229	0.740271	0.292634	0.292651	0.115687	0.115693	0.045734	0.045737
4	8	0.677177	0.677195	0.275521	0.275529	0.112101	0.112104	0.045610	0.045611
5	12	0.596276	0.596246	0.252644	0.252632	0.107046	0.107041	0.045356	0.045354
6	14	0.567794	0.56753	0.244314	0.244199	0.105124	0.105075	0.045234	0.045212
7	16	0.544078	0.544249	0.237256	0.23733	0.103460	0.103492	0.045116	0.04513
8	18	0.523869	0.52395	0.231150	0.231186	0.101992	0.102008	0.045002	0.045009
9	22	0.490917	0.490493	0.221001	0.220809	0.099490	0.099404	0.044788	0.044749

A5. The function $x = f(\dot{\kappa})$

The function *lambda* acts on s with additional parameters $x \in \mathbb{R}$ and l with $0 < x \leq 1$ so that

$$\lambda_l(s, x) = \ell_{l,\kappa-1}(s) + x\vec{\mathcal{L}}_\kappa - \left(h_{l,\kappa-1}(s) + x\vec{\mathcal{R}}_\kappa\right).$$

This exercise determined x from the intersection of $\vec{\mathcal{L}}_\kappa$ and $\vec{\mathcal{R}}_\kappa$ at known zeros, this is formally the same as determining x to satisfy for each ρ_i

$$\ell_{l,\kappa-1}(\rho_i) + x\vec{\mathcal{L}}_\kappa = h_{l,\kappa-1}(\rho_i) + x\vec{\mathcal{R}}_\kappa.$$

In Figure A4 x is plotted against t for all the t_i in an interval in t.

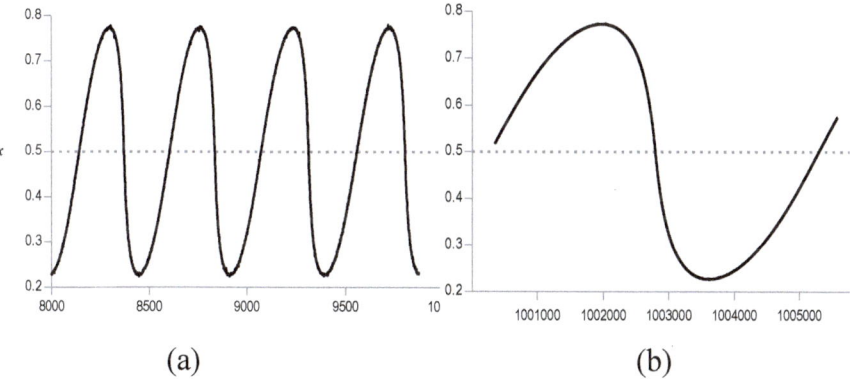

(a)   (b)

oscillates about $x = 1/2$ with a decreasing frequency.

Figure A4 shows the frequency of oscillation relates to t and Figure A5 adds in a plot for $\dot{\kappa}$ as a function of t.

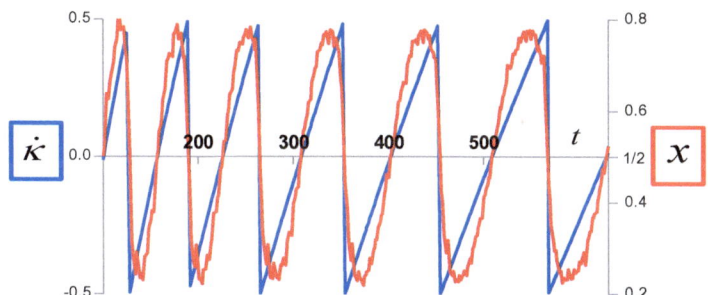

Figure A5. In red, x on the right-hand axis representing the intersection of $\vec{\mathcal{L}}_\kappa$ and $\vec{\mathcal{R}}_\kappa$ for all the *zeros* in the interval $t = 100$ to $t = 630$ and in blue, $\dot{\kappa}$ on the left-hand axis.

The relationship between x and $\dot{\kappa}$ without t appears in Figure A6.

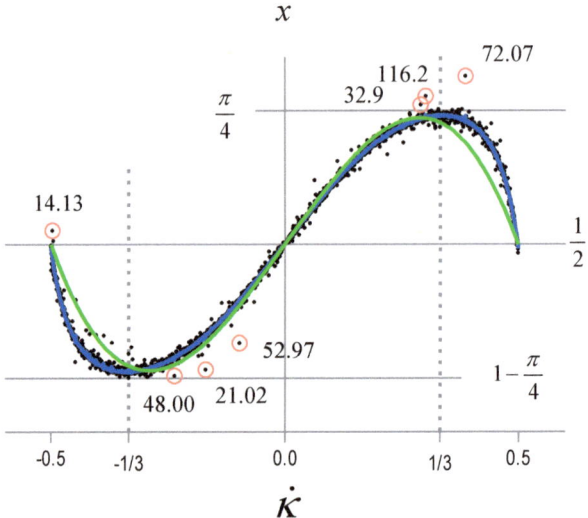

Figure A6. The first 10,000 *zeros* illustrate the relationship between $\dot{\kappa}$ (abscissa) and x (ordinate). A smoothed curve is plotted in blue and it is noted that the principal outliers, encircled in red, all have low values of t. In green is a crude approximation based on the first two terms of the Taylor series for a sine function.

Figure A6 uses $l = \kappa + 1$; increasing l further will not improve this plot. Kappa dot is limited to $-1/2 < \dot{\kappa} < 1/2$ and $0 < x \leq 1$. The value x appears to have bounds of $\sim\pi/4$ and $\sim(1 - \pi/4)$ which occur at $\dot{\kappa}\sim 1/3$ and $\dot{\kappa}\sim -1/3$ respectively. These bounds have not been examined in detail and are merely observations. The green curve in Figure A6 is a crude approximation based on a transformation using the first two terms of the Taylor series for the sine function $\theta - \theta^3/3!$

$$x \sim \frac{14\pi\dot{\kappa}}{9} - \left(\frac{\left(\frac{14\pi\dot{\kappa}}{9}\right)^3}{3!}\right)\left(\frac{\pi}{4} - \frac{1}{2}\right) + \frac{1}{2}.$$

Figure A7 is an example of an intersection of *kappa* vectors when $\dot{\kappa} \approx 0$.

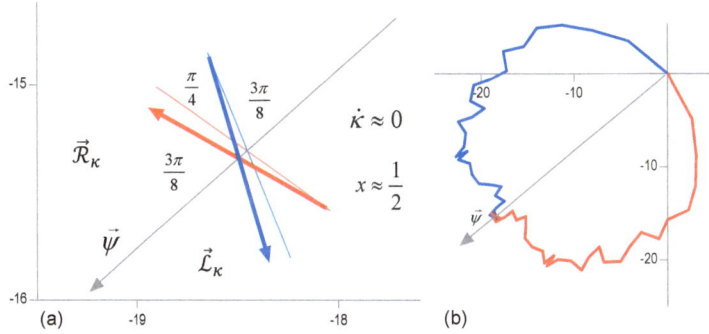

Figure A7. (a) An illustration of $\vec{\mathcal{R}}_\kappa$ from $\mathsf{P}(h_{l,\kappa}(s))$ in red and $\vec{\mathcal{L}}_\kappa$ from $\mathsf{P}(\ell_{l,\kappa}(s))$ in blue for the *zero* $t_i = 5655.819253$ at $\sigma = 1/2$ with $\kappa = 30$ and $l = 31$. (b) The two pathways whose final vectors are shown in (a). The figure shows that if $\dot{\kappa} \approx 0$ then $x \approx 1/2$ and the vectors $\vec{\mathcal{R}}_\kappa$ and $\vec{\mathcal{L}}_\kappa$ intersect at about $\pi/4$.

Figure A8 shows the angle between ψ_0 and $\arg(\vec{\mathcal{L}}_\kappa)$ plotted (a) against x and (b) against $\dot{\kappa} \approx 0$.

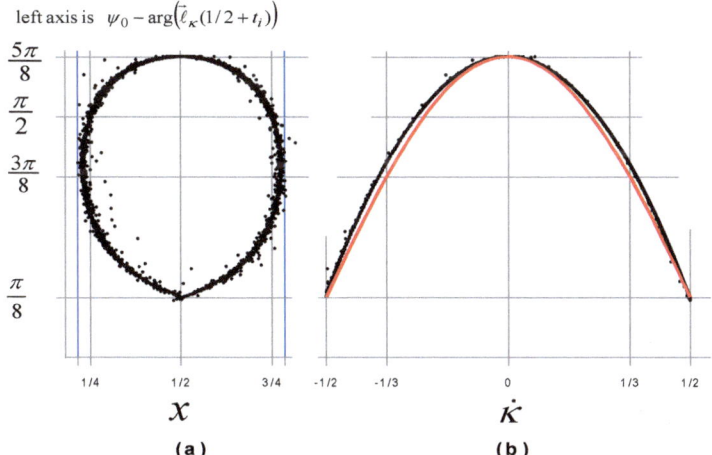

Figure A8. Both figures are dot plots for the first 10,000 *zeros*. The vertical axis in both figures is the reduced argument $\psi_0 - \arg(\vec{\mathcal{L}}_\kappa)$. In (a) the horizontal axis is x with the blue lines at $x = \pi/4$ and $x = 1 - \pi/4$. In (b) the horizontal axis is $\dot{\kappa}$. The difference in arguments is roughly tracked by the red cosine curve with amplitude $\pi/2$, period $2\dot{\kappa}$ and added to $\pi/8$.

If at a *zero* the difference in arguments were exactly $\pi/2$ then the $\vec{\ell}_\kappa$ vector and the $\vec{\mathcal{R}}_\kappa$ vector would be collinear and x would not be defined.

An example of a nontrivial zero with nearly collinear kappa vectors

In Figure A9 a *zero* with $l = \kappa + 1$ is illustrated with the *kappa* vectors nearly collinear to each other where they meet.

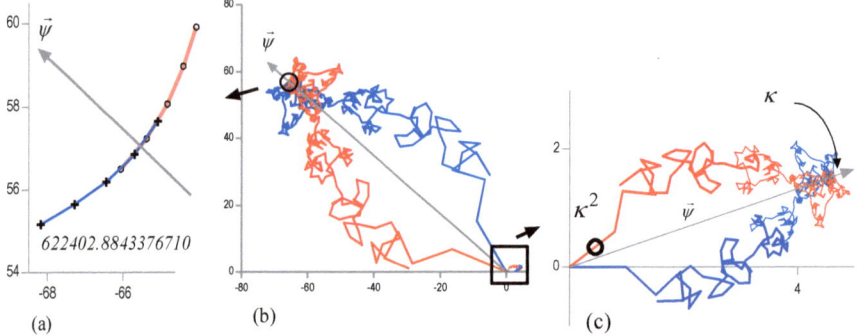

Figure A9. $\mathbf{P}(h_{l,\kappa}(s))$ in red and $\mathbf{P}(\ell_{l,\kappa}(s))$ in blue with $\kappa = 315$ and $l = 316$ for $t_i = 622402.884$ at $\sigma = 1/2$. **(a)** The *kappa* vectors are almost collinear at 92.7 degrees to $\vec{\psi}$. **(b)** Both pathways in full; the bottom right-hand corner shows the associated $\mathbf{P}(\zeta_{\kappa l}(s))$ which nearly vanishes at the *zero*. **(c)** The partial Euler's *zeta*, is taken from (b) and enlarged with $\mathbf{P}(h_{1,\kappa}(s))$ in red and $\mathbf{P}(\zeta_{\kappa}(s))$ in blue. The black circle represents the end of the partial series $\mathrm{P}(\zeta_{\kappa l}(s))$, with $\vec{m}$ for $m = \kappa l$ near the *inflection point* in the $\mathrm{P}(\mathcal{R}_1)$ of $\mathrm{P}(\zeta_{\tau}(s))$.

It is likely that true collinearity could not occur for any t_i. Figure A9 also illustrates $\zeta_\tau(\rho) \approx 0$ when $l = \kappa + 1$.

An intricate pathway and kappa vectors at an extreme angle

In Figure A10 (below) the *kappa* vectors lie in a similar direction which may seem counter intuitive.

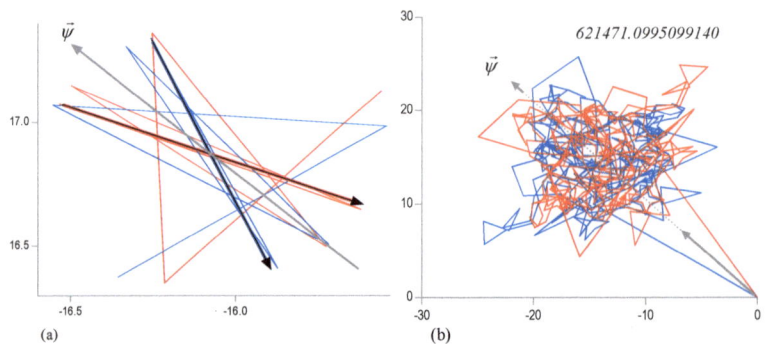

Figure A10. A nontrivial zero with *kappa* vectors meeting at 22.5 degrees; $\kappa = 314$, $t = 621471.099509914$.

A6. An example showing failure of superposition either side of a zero and showing "opportunistic superstructures" in the proximal pathway.

In Figure A11 a short region of the mirror image pathways for values of t slightly below a t_i in (a), for the t_i in (b) and for a t slightly above the t_i in (c). In (b) there is symmetry and perfect superposition at the zero either side of $\vec{\psi}$ such that the light blue dashed line of $P(\ell_{l,n}(s))$ beyond κ terms overlies $P(h_{l,\kappa}(s))$, but not when $t \neq t_i$.

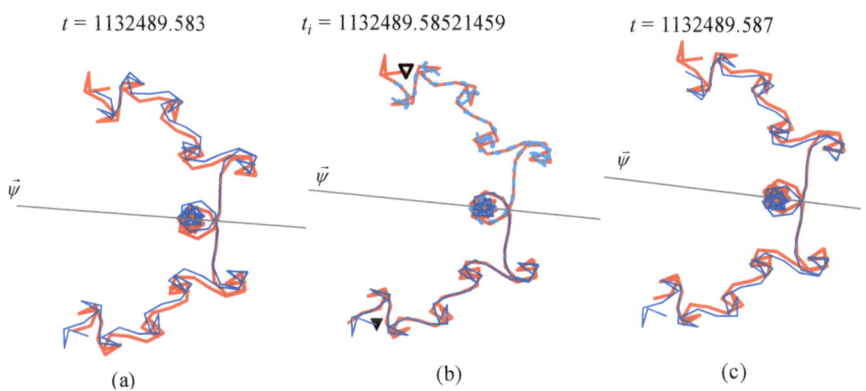

Figure A11. Three plots of parts of two pathways. In **(a)** $t < t_i$ in **(b)** $t = t_i$ with i being 200,1050, and in **(c)** $t > t_i$. $P(\ell_{l,n}(s))$ is shown in blue, starting at the bottom of (b) with a solid black triangle. $P(h_{l,\kappa}(s))$ is in red, starting at the top of (b) with an open triangle. Each pathway has been extended beyond their *kappa* vector to show the congruence of the pathways.

In Figure A11 there appear to be $P(\mathcal{R}_r)$-like superstructures in $P(\ell_{l,n}(s))$ and $P(h_{l,r}(s))$ before their *kappa* vectors. This is simply because sequential $\mathrm{Arg}\left(t\ln\left(\frac{n+1}{n}\right)\right)$, which is the reduced form, generate a closely falling sequence, however since $t\ln\left(\frac{(n+1)^2}{n(n+2)}\right) > 2\pi$ these are not true superstructures. It is important to appreciate that if

$$t\ln\left(\frac{n+1}{n}\right) - t\ln\left(\frac{n+2}{n+1}\right) > 2\pi$$

the region is in the *proximal* pathway. This analysis is formally the same as noting that if two l^{th} vectors differ in argument by $t\ln\left(\frac{(k+1)l}{kl}\right) \cong 2\pi$ then $kl \cong \left(e^{2\pi/t} - 1\right)^{-1}$ and with $l = \kappa + 1$ we have $\kappa \cong \left(e^{2\pi/t} - 1\right)^{-1/2}$.

A7. The $P(\mathcal{M}_m)$ in the distal $P(\bar{h}_{l,r}(s))$ and the $P(L_n)$ structures in the distal $P(\ell_{l,n}(s))$

A proof of RH only requires $P(h_{l,\kappa}(s))$ when $l > \kappa$. However, it would be a lost opportunity not to illustrate the *distal* $P(\bar{h}_{l,r}(s))$ when $r > \kappa$, and so the following illustrates $l < \kappa$, with $t = 1230.58$, $\kappa = 14$ and $l = 4$.

A $P(\mathcal{M}_m)$ is a coherent set $\mathcal{M}_m$ of $\vec{\mathcal{R}}_r$ in the *distal* $P(\bar{h}_{l,r}(s))$: $\mathcal{M}_m \subseteq \{\vec{\mathcal{R}}_r\}$. An integer ϱ (*rho variant*) determines each $q_{\tilde{r}}$ that identifies an important $\vec{\mathcal{R}}_r$, in a $P(\mathcal{M}_m)$. The integer ϱ has subscripts, m or $\bar{m}$, with $\varrho_m \in \mathbb{O}$ referring to a pseudo-convergence, and $\varrho_{\bar{m}} \in \mathbb{E}$ referring to an *inflection point*. This is akin to $q \in \mathbb{N}$ determining an m which identifies an important $\vec{m}$, and having $q_r \in R_l$ referring to a pseudo-convergence, and $q_{\tilde{r}} \in \chi_l$ referring to an *inflection point* in a $P(\mathcal{R}_r)$. Unlike $\vec{\mathcal{R}}_r$ the vector $\vec{m}$ needs no subscript and the ordered sets $\mathbb{O}$ and $\mathbb{E}_0$, unlike R_l and χ_l, are independent of l. The final pseudo-convergence of $P(\bar{h}_{l,r}(s))$ has a point $\bar{h}_{l,r}(s)$ at $r = \iota$ at the start of divergence.

The focal point of the pseudo-convergences and the *inflection point* in each $P(\mathcal{M}_m)$ are generated by the diminishing $\vec{\mathcal{R}}_r$ vectors of the *distal* $P(\bar{h}_{l,r}(s))$ when $r > \kappa$. These points, located to a $q_{\tilde{r}}$, occur when the arguments of neighbouring $\vec{\mathcal{R}}_r$ vectors differ by an amount that is closest to a specific integral multiple of π/l, with that integer being ϱ;

$$\frac{\varrho\pi}{l} \cong t\ln\left(\frac{q_{\tilde{r}}+2}{q_{\tilde{r}}}\right) \text{ with } \varrho \in \mathbb{Z}. \tag{131}$$

Equation 131 follows from the geometry of $P(\bar{h}_{l,r}(s))$ and allows the integer ϱ to specify significant vectors. Sequential $q_{\tilde{r}}$, rise in χ_l in increments of 2 or 4. For convenience we first assume the rise is 2 and then determine the nearest element of the set, either χ_l or R_l, using the brackets $[\![x]\!]$ to mean the nearest element of the appropriate set. The inequality

$$\frac{tl}{\pi}\ln\left(\frac{q_{\tilde{r}}+4}{q_{\tilde{r}}+2}\right) < \varrho \leq \frac{tl}{\pi}\ln\left(\frac{q_{\tilde{r}}+2}{q_{\tilde{r}}}\right) \tag{132}$$

follows from the geometry of the l sided polygons. The *inflection point* in the $P(\mathcal{M}_m)$ for each m lies at a specific $q_{\tilde{r}}$ as determined by Equation 133 below. This point lies very close to the mid-point of the $\vec{m}$ which equates to the *principal-axis* of the $P(\mathcal{M}_m)$. The integral multiple of $\pi/(tl)$ is designated $\varrho_{\bar{m}}$ (analogous to $q_{\tilde{r}}$) with the bar indicating the association with the *inflection point*. If $q_{\bar{r}}$ were allowed to take any real value then $t\ln((q_{\bar{r}} + 2)/q_{\bar{r}}) = \pi\varrho_m/l$ but the nearest we can establish will be to choose $q_{\bar{r}} \in \chi_l$ as in;

$$q_{\bar{r}} = \left[\left(e^{\frac{\pi\varrho_{\bar{m}}}{tl}} - 1\right)^{-1} + \left(e^{\frac{-\pi\varrho_{\bar{m}}}{tl}} - 1\right)^{-1}\right] \text{ with } \varrho_{\bar{m}} \in \mathbb{E}_2 \quad (133)$$

$$\text{and } q_{\bar{r}} = \left[\left(e^{\frac{\pi\varrho_m}{tl}} - 1\right)^{-1} + \left(e^{\frac{-\pi\varrho_m}{tl}} - 1\right)^{-1}\right] \text{ with } \varrho_m \in \mathbb{O}. \quad (134)$$

It is important to be clear that each $q_{\bar{r}}$ relates to the mid-part of a $\vec{\mathcal{R}}_r$ and a pair of q_r (without the bar) represent the ends of the $\vec{\mathcal{R}}_r$. Here we are looking at the $\vec{m}$ whose centres are related to $\varrho_{\bar{m}}$, each determined from a $q_{\bar{r}}$, and whose ends are related to the ϱ_m which are also determined from a $q_{\bar{r}}$. This is illustrated in Figure A12(a).

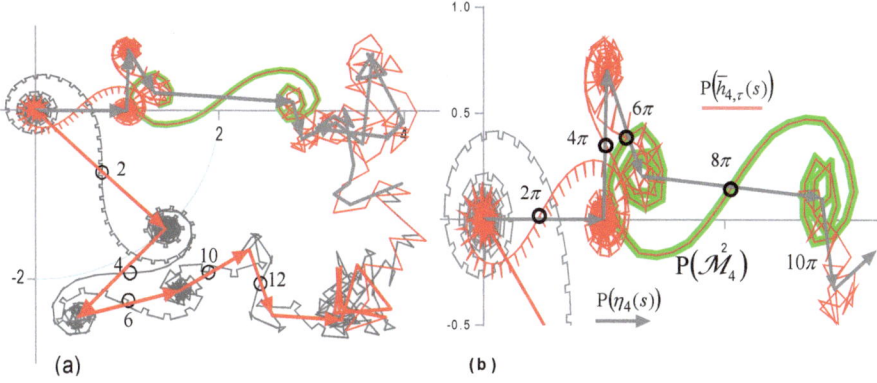

Figure A12. (a) $P(\eta_4(s))$ in grey for $t = 1230.5846$, $\sigma = 1/2$. In red, the $\vec{\mathcal{R}}_r$ of $P(\bar{h}_{l,n}(s))$ retrace the $P(\mathcal{R}_r)$ of the *distal* $P(\eta_{4,n}(s))$ as $q_{\bar{r}}$ runs $\{2, 4, 6, 10, 12 \ldots\}$, $8 \notin \chi_4$. Overlying the *proximal* $P(\eta_{4,\tau}(s))$ the many small $\vec{\mathcal{R}}_r$ of the *distal* $P(\bar{h}_{l,n}(s))$, in red, form the $P(\mathcal{M}_m)$. (b) Detail showing the *inflection points* in $P(\mathcal{M}_1)$ to $P(\mathcal{M}_4)$, the later highlighted in green.

In Figure A12(b), the $\vec{m}$ for $m = 4$ has a magnitude of $3/\sqrt{4}$ and is associated with the largest superstructure $P(\mathcal{M}_4)$ which is highlighted in green. The $P(\mathcal{M}_4)$ structure has a number of $\vec{\mathcal{R}}_r$ falling into a uniform progression at around 8π, in contrast to the three subsequent $P(\mathcal{M}_m)$ where 2 in 4 spike to one side of the pathway's progression at 6π, and where 1 in 4 double-back on the pathway's progression at 4π, see Figure A13 overleaf, in this way $P(\mathcal{M}_4)$ achieves its greater size.

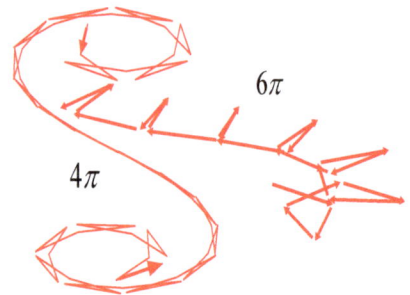

Figure A13. Detail of the geometry at 6π and 4π from $P(\bar{h}_{4,r}(s))$ taken from Figure A12(b).

In Figure A12 (page 117) the superposition of the $P(\mathcal{M}_m)$ (made of many $\vec{\mathcal{R}}_r$) upon the $\vec{m}$ is clear. Importantly, the figure shows the relationship between $\vec{4}$ having a larger magnitude than its neighbours and the mechanism by which the $P(\mathcal{M}_m)$ is able to match this change in magnitude through simple geometry. In Table A7 the relationship between m, ϱ_m, $q_{\bar{r}}$ and r is detailed. To be explicit, an m specifies a ϱ_m since $m = \varrho_m/2$, then Equation 134 gives $q_{\bar{r}} = f(\varrho_m)$ and then from χ_l the $q_{\bar{r}}$ yields the value r by Equation 42 (page 23), which is its position in the ordered set.

Table A7. Detail from Figure A12. The relationship of r in an $\vec{\mathcal{R}}_r$ to the inflection point in the $P(\mathcal{M}_m)$.

Inflection in $P(\mathcal{M}_m)$	$\varrho_{\bar{m}}$	Figure A12(b)	Content of Equation 1	$q_{\bar{r}} \in \chi_4$	$\vec{\mathcal{R}}_r$ r
$P(\mathcal{M}_1)$	2	2π	1566.83	1566	588
$P(\mathcal{M}_2)$	4	4π	783.41	782	294
$P(\mathcal{M}_3)$	6	6π	522.28	522	196
$P(\mathcal{M}_4)$	8	8π	391.71	390	147

The final pseudo-convergence in $P(\mathcal{M}_1)$ from Equation 134 is at $q_{\bar{r}} = 3134$ which relates to $r = 1176$ by Equation 42 (page 23) as shown in Figure A14.

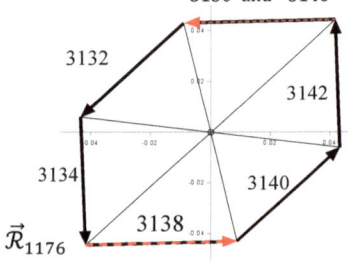

Figure A14. Final pseudo-convergence of $P(\bar{h}_{4,r}(s))$ in the region specified by $\varrho_m = 1$ with $q_{\bar{r}} = [\![3133.65]\!] = 3134$ which locates $\vec{\mathcal{R}}_{1176}$. This vector with its 5 neighbours provides an average remarkably close to zero.

In like manner the *distal* part of $P\big(\ell_{l,\mathfrak{n}}(s)\big)$ is illustrated with $\mathfrak{n} = \tau$ in Figure A15.

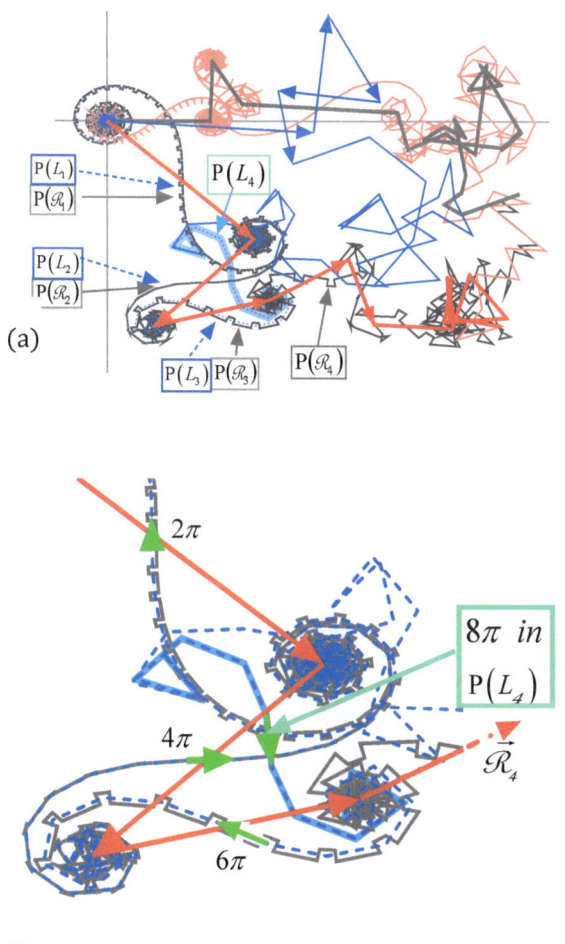

Figure A15. (a) With $t_{839} = 1230.5846$ and $\sigma = 1/2$, $P(\eta_4(s))$ is shown in grey/black, and in red $P\big(\overline{h}_{4,\tau}(s)\big)$ with its early vectors emboldened. In blue $P\big(\ell_{4,\tau}(s)\big)$ is shown and this would be complimentary to $P(h_{4,\tau}(s))$ which is not shown. The first three $P(L_n)$ are superimposed on the first three $P(\mathcal{R}_r)$ and relate to the first three $\vec{\mathcal{R}}_r$ vectors, but the next two do not share locations. **(b)** Detail from (a) showing the associated *inflection points* in $P\big(\ell_{4,\tau}(s)\big)$ crossing the first three $\vec{\mathcal{R}}_r$ in red. Sequential $\vec{L}_n$ near the *inflection*

points have arguments that differ by multiples of 2π. There is no superposition of the P(L_4) with $\vec{\mathcal{R}}_4$ at 8π and of course $8 \notin \chi_4$.

This section illustrated the mechanics of superposition of *distal* superstructures of one pathway upon the *proximal* vectors of another to show how these can be complimentary.

A8. Observed and expected number of intervals containing n zeros

Table A8. The Observed and Expected values if a Poisson model were to apply.

Number of zeros in an interval	Number of Intervals Observed	Number of zeros	Number of Intervals Poisson	Number of zeros expected
0			67.89	0
1			251.71	252
2	50	100	466.63	933
3	947	2841	576.69	1730
4	1533	6132	534.54	2138
5	236	1180	396.37	1982
6	1	6	244.93	1470
7			129.73	908
8			60.12	481
9			24.77	223
10			9.18	92
11			3.10	34
12			0.96	11
13			0.27	4
14			0.07	1
Total	2767	10259	2767	10259

This table's figures populate Figure 55 on page 85. The Poisson model used $\lambda = 3.707$ for 10259 zeros and 2767 intervals.

A9. Zeros of beta

Table A9. Estimated zeros of $\beta(1/2 + it)$: first 320

Based on the minima of $\beta_\tau(s)$ or at very low values of t using $\beta_{2\tau}(s)$.

#	value	#	value	#	value	#	value	#	value	#	value	#	value	#	value
1	6.0114	41	84.732	81	143.33	121	197.115	161	247.996	201	296.902	241	344.085	281	390.17
2	10.264	42	86.578	82	144.979	122	198.808	162	249.185	202	298.08	242	345.211	282	391.083
3	13.002	43	87.631	83	146.523	123	200.163	163	251.087	203	298.871	243	346.264	283	392.853
4	16.353	44	89.802	84	147.935	124	200.903	164	251.638	204	300.436	244	347.926	284	393.444
5	18.285	45	91.351	85	149.189	125	202.261	165	252.625	205	301.921	245	348.983	285	394.821
6	21.445	46	92.239	86	150.298	126	204.222	166	254.315	206	302.919	246	349.435	286	395.806
7	23.274	47	94.168	87	151.963	127	204.993	167	255.839	207	303.662	247	350.978	287	396.736
8	25.726	48	96.138	88	153.701	128	206.413	168	256.506	208	305.066	248	352.393	288	398.251
9	28.358	49	96.963	89	154.577	129	207.318	169	258.166	209	306.796	249	353.694	289	399.566
10	29.662	50	98.756	90	155.651	130	209.229	170	258.836	210	307.287	250	354.303	290	400.694
11	32.596	51	100.136	91	157.749	131	210.104	171	260.432	211	309.123	251	355.745	291	400.827
12	34.204	52	102.142	92	158.706	132	211.834	172	261.915	212	309.741	252	356.629	292	402.819
13	36.145	53	103.288	93	160.238	133	212.539	173	262.885	213	310.825	253	358.289	293	403.908
14	38.512	54	104.334	94	161.408	134	213.763	174	264.055	214	312.557	254	359.098	294	404.947
15	40.321	55	106.695	95	162.567	135	215.795	175	264.934	215	313.879	255	360.712	295	406.162
16	41.806	56	107.691	96	164.732	136	216.705	176	267.002	216	314.434	256	361.199	296	407.146
17	44.618	57	109.261	97	165.402	137	217.583	177	267.802	217	315.736	257	362.278	297	408.04
18	45.599	58	110.501	98	166.755	138	219.174	178	268.784	218	317.013	258	364.421	298	409.589
19	47.742	59	112.369	99	168.045	139	220.407	179	270.279	219	318.457	259	364.805	299	410.873
20	49.725	60	113.816	100	170.052	140	221.93	180	271.256	220	319.581	260	366.051	300	411.662
21	51.688	61	115.144	101	170.736	141	223.004	181	272.757	221	320.407	261	367.126	301	412.732
22	52.771	62	116.194	102	172.282	142	224.123	182	274.172	222	322.003	262	368.435	302	413.67
23	55.27	63	118.539	103	173.444	143	225.294	183	275.034	223	322.54	263	369.503	303	415.43
24	56.937	64	119.454	104	174.916	144	226.989	184	275.86	224	324.519	264	371.07	304	416.372
25	58.117	65	120.732	105	176.598	145	228.406	185	277.773	225	325.476	265	371.771	305	417.13
26	60.423	66	122.448	106	177.703	146	228.96	186	278.805	226	326.459	266	372.886	306	418.556
27	62.009	67	123.795	107	178.363	147	230.331	187	280.158	227	327.261	267	373.936	307	419.488
28	63.714	68	125.769	108	180.57	148	232.101	188	280.794	228	329.201	268	375.548	308	420.53
29	64.976	69	126.299	109	181.616	149	233.049	189	282.381	229	330.037	269	376.685	309	422.511
30	67.636	70	127.96	110	182.918	150	234.353	190	283.606	230	331.212	270	377.56	310	422.688
31	68.369	71	129.886	111	184.116	151	235.837	191	284.927	231	332.574	271	378.388	311	424.068
32	70.188	72	131.094	112	185.375	152	236.243	192	286.081	232	333.183	272	380.058	312	424.824
33	72.157	73	132.145	113	187.07	153	238.538	193	287.15	233	334.77	273	381.05	313	426.733
34	73.77	74	133.745	114	188.272	154	239.34	194	287.979	234	336.154	274	382.241	314	427.385
35	75.145	75	135.492	115	189.493	155	240.628	195	290.253	235	337.141	275	383.486	315	428.699
36	76.697	76	136.549	116	190.372	156	241.479	196	290.678	236	338.227	276	384.374	316	429.59
37	78.812	77	138.459	117	192.362	157	243.23	197	291.833	237	339.012	277	385.216	317	430.619
38	80.211	78	138.751	118	193.798	158	244.514	198	293.203	238	340.843	278	387.15	318	432.062
39	81.214	79	141.255	119	194.233	159	245.566	199	294.328	239	342.027	279	388.083	319	433.038
40	83.668	80	142.395	120	196.133	160	246.725	200	295.804	240	342.687	280	388.779	320	434.44

Table A9. Estimated zeros of $\beta(1/2 + it)$: continued

5628th – 5947th zero of $\beta(1/2 + it)$

#	value	#	value	#	value	#	value	#	value	#	value	#	value	#	value
5628	5003.079	5668	5033.827	5708	5065.304	5748	5096.129	5788	5127.597	5828	5158.475	5868	5188.363	5908	5219.86
5629	5003.774	5669	5034.605	5709	5066.182	5749	5096.94	5789	5128.243	5829	5159.317	5869	5189.502	5909	5220.628
5630	5004.344	5670	5035.485	5710	5066.487	5750	5097.597	5790	5128.632	5830	5159.98	5870	5190.078	5910	5221.493
5631	5005.238	5671	5036.664	5711	5067.555	5751	5098.659	5791	5129.999	5831	5160.061	5871	5191.196	5911	5221.8
5632	5005.889	5672	5037.438	5712	5068.292	5752	5099.555	5792	5130.6	5832	5161.903	5872	5191.789	5912	5222.354
5633	5006.732	5673	5037.902	5713	5068.986	5753	5100.233	5793	5130.919	5833	5162.282	5873	5192.59	5913	5223.726
5634	5007.824	5674	5038.593	5714	5069.822	5754	5101.194	5794	5131.909	5834	5163.298	5874	5193.311	5914	5224.151
5635	5008.225	5675	5039.509	5715	5070.402	5755	5101.643	5795	5132.658	5835	5163.708	5875	5194.172	5915	5225.345
5636	5009.26	5676	5040.114	5716	5071.856	5756	5102.113	5796	5133.758	5836	5164.499	5876	5194.574	5916	5225.847
5637	5010.155	5677	5041.019	5717	5072.247	5757	5103.312	5797	5134.422	5837	5165.451	5877	5195.505	5917	5226.647
5638	5010.689	5678	5041.959	5718	5073.039	5758	5104.285	5798	5135.308	5838	5166.148	5878	5196.496	5918	5227.517
5639	5011.169	5679	5042.727	5719	5073.416	5759	5104.564	5799	5136.007	5839	5166.775	5879	5197.467	5919	5228.157
5640	5012.031	5680	5043.387	5720	5074.387	5760	5105.802	5800	5136.293	5840	5167.702	5880	5198.146	5920	5228.722
5641	5013.165	5681	5044.435	5721	5075.208	5761	5106.364	5801	5137.484	5841	5168.583	5881	5198.682	5921	5229.771
5642	5014.05	5682	5044.633	5722	5076.285	5762	5107.281	5802	5137.998	5842	5169.591	5882	5199.228	5922	5230.472
5643	5014.859	5683	5045.878	5723	5076.874	5763	5108.065	5803	5139.19	5843	5170.166	5883	5200.135	5923	5231.33
5644	5015.218	5684	5046.333	5724	5077.761	5764	5108.507	5804	5139.99	5844	5170.445	5884	5201.272	5924	5232.341
5645	5015.868	5685	5047.061	5725	5078.333	5765	5109.101	5805	5140.6	5845	5171.569	5885	5201.869	5925	5232.896
5646	5017.11	5686	5048.281	5726	5079.172	5766	5110.407	5806	5141.126	5846	5172.294	5886	5202.466	5926	5233.518
5647	5017.676	5687	5049.153	5727	5079.872	5767	5111.402	5807	5142.159	5847	5173.002	5887	5203.667	5927	5234.274
5648	5018.271	5688	5049.713	5728	5080.582	5768	5111.838	5808	5143.003	5848	5174.3	5888	5204.127	5928	5234.82
5649	5019.328	5689	5050.223	5729	5081.652	5769	5112.541	5809	5143.524	5849	5174.867	5889	5205.013	5929	5236.124
5650	5020.319	5690	5050.9	5730	5082.151	5770	5113.276	5810	5144.116	5850	5175.342	5890	5205.596	5930	5237.024
5651	5020.81	5691	5051.944	5731	5083.365	5771	5114.079	5811	5145.445	5851	5176.109	5891	5206.353	5931	5237.397
5652	5021.631	5692	5052.777	5732	5083.812	5772	5114.925	5812	5146.251	5852	5177.057	5892	5207.032	5932	5238.382
5653	5022.308	5693	5053.694	5733	5084.727	5773	5115.545	5813	5146.851	5853	5177.645	5893	5208.476	5933	5238.722
5654	5023.08	5694	5054.247	5734	5085.344	5774	5116.675	5814	5147.42	5854	5178.608	5894	5208.897	5934	5240.101
5655	5023.659	5695	5055.148	5735	5085.877	5775	5117.46	5815	5148.193	5855	5179.523	5895	5209.488	5935	5240.486
5656	5024.88	5696	5055.772	5736	5086.735	5776	5117.86	5816	5148.84	5856	5179.975	5896	5210.536	5936	5241.043
5657	5025.87	5697	5056.869	5737	5088.268	5777	5118.964	5817	5149.98	5857	5181.08	5897	5210.899	5937	5242.142
5658	5026.143	5698	5057.257	5738	5088.555	5778	5119.515	5818	5150.819	5858	5181.919	5898	5211.822	5938	5243.269
5659	5027.115	5699	5057.724	5739	5089.225	5779	5120.307	5819	5151.403	5859	5182.326	5899	5212.504	5939	5243.782
5660	5027.863	5700	5059.315	5740	5089.858	5780	5120.904	5820	5152.301	5860	5182.987	5900	5213.681	5940	5244.28
5661	5028.186	5701	5059.79	5741	5090.929	5781	5121.884	5821	5152.961	5861	5183.704	5901	5214.271	5941	5245.395
5662	5029.543	5702	5060.47	5742	5091.752	5782	5123.076	5822	5153.701	5862	5184.413	5902	5214.886	5942	5245.736
5663	5030.183	5703	5061.641	5743	5092.007	5783	5123.662	5823	5154.457	5863	5185.193	5903	5215.76	5943	5246.51
5664	5031.175	5704	5061.867	5744	5093.33	5784	5124.072	5824	5155.155	5864	5185.64	5904	5216.832	5944	5247.848
5665	5031.43	5705	5062.618	5745	5093.914	5785	5124.999	5825	5156.123	5865	5186.412	5905	5217.155	5945	5248.38
5666	5032.81	5706	5063.423	5746	5094.889	5786	5125.523	5826	5156.941	5866	5186.944	5906	5217.903	5946	5249.191
5667	5033.405	5707	5064.601	5747	5095.72	5787	5126.545	5827	5157.74	5867	5188.042	5907	5218.797	5947	5249.924

Notation

α, β	reserved for σ_α and σ_β when $0 < \sigma_\alpha < 1/2 < \sigma_\beta < 1$
γ	the domain of *upsilon* $\Upsilon(\gamma)$ a smooth spiral curve
$\zeta(s)$	reserved for Riemann's *zeta* function valid for all s
$\zeta_n(s)$	partial Euler's *zeta* function to n terms
$\eta(s)$	Dirichlet's *eta* function — an alternating Euler's *zeta* function
$\eta_l(s)$	modified *eta* function: every l^{th} term of *zeta* is multiplied by $1-l$
$\eta_{l,n}(s)$	modified partial *eta* series to n terms
$\eta_l(\sigma)$	$\eta_l(s)$ with t fixed and considered as an arc over an interval in σ
$\eta_l(t)$	$\eta_l(s)$ with σ fixed and considered as an arc over an interval in t
$\hat{\eta}_q(s)$	represents the focal point of a pseudo-convergence for a stated l
$\Delta\theta$	the reduced argument between neighbouring vectors
$\Delta\vartheta$	the primitive argument between neighbouring vectors
κ and $\dot{\kappa}$	*Kappa* and *kappa* dot, integer and fraction relating to the intersection of vectors at the junction of *proximal* and *distal* pathways
$\lambda_l(s,x)$	difference between two paired series
$\mu(n)$	the Möbius function, $\mu \in \mathbb{R}$ is also used as a dummy variable
ρ	a nontrivial zero of the Riemann *zeta* function $\{\rho\} = \{\rho_i\} + \{\rho_u\}$
ρ_i	known nontrivial zero on the *critical line* with $\sigma = 1/2$
ρ_u	un-known (hypothetical) zero off the *critical line* with $\sigma \neq 1/2$
ϱ	*Rho* variant is used as ϱ_m and $\varrho_{\bar{m}}$ to determine a value of $q_{\bar{r}}$
σ	Re(s) the real part of the domain of *zeta, eta* etc.,
σ_a, σ_b	values of σ such that $\sigma_a < \sigma_b$ but without restrictions
σ_c	σ at a cycloid-like curve where the differential will be zero
τ	an integer whose vector $\vec{\tau}$ precedes the spiral of divergence
$\Upsilon(\gamma)$	*Upsilon* a smooth curve following paired-spirals
$\phi(s)$	*Phi* variant locates the focal point of a pseudo-convergence
χ_l	the set $\{q_{\bar{1}}, q_{\bar{2}}, q_{\bar{3}}, \dots q_{\bar{r}}, \dots\}$ identifying *inflection points*
$\vec{\psi}$	the line of reflection, *Psi* $\psi = \arg(\vec{\psi})$
ψ_0, ψ_π	arguments of the rays $\vec{\psi}_0$ and $\vec{\psi}_\pi$
ψ_i	the ψ_0 associated with a ρ_i
$\arg(z)$	unreduced argument of a complex number: $\arg(z) \in \mathbb{R}$
$\text{Arg}(z)$	principal value of $\arg(z)$ as $[0, 2\pi)$ or $(-\pi, \pi]$

$h_{l,r}(s)$	a vector series summating $\vec{\mathcal{R}}_r$
i	either $\sqrt{-1}$ or the index in $t_i,\ p_i,\ \rho_i,\ \psi_i,\ \nu_i$
j	$j \in \mathbb{N}_1$ used in the definition of $\eta_l(s)$ in two ways
k	Roman k, an integer $k \in \mathbb{E}$ as in $t = 2\pi k/\ln(l)$
l	$l \in \mathbb{N}_1$ in $\eta_l(s)$ when $l \geq 2$ and $l = 1$ in Euler's zeta
$\mathcal{M}_m$,	a set of $\vec{\mathcal{R}}_r$ forming paired pseudo-spirals
$\vec{m}$	vector with index m in Euler's zeta, $\eta_{l,n}(s)$ or $\beta_n(s)$
$|\vec{m}|$	the magnitude of the vector $\vec{m}$
m, n	$m, n \in \mathbb{N}_1$ used for terms in Euler's zeta, $\eta_{l,n}(s)$ or $\beta_n(s)$
p_i	an indexed prime
$P\big(f_{l,n}(s)\big)$	the pathway of vectors in a series $f(s)$
$q,\ q_r, q_{\bar{r}}$	integral multipliers of π/t or $\pi/(tl)$ referring to parts of a $P(\mathcal{R}_r)$
r	the index of $P(\mathcal{R}_r)$ and $P(\mathcal{R}'_r)$
$\bar{r}$	r carries a bar when it relates to the *inflection point* in a $P(\mathcal{R}_r)$
R_l	$R_l = \{q_r\}$ identifying pseudo-convergences
$\mathcal{R}_r$	the set of vectors in a $P(\mathcal{R}_r)$ in $P(\zeta_n(s))$, $P(\eta_l(s))$ or $P(\beta_n(s))$
$\vec{\mathcal{R}}_r$	a vector sharing properties with the *principal-axis* of a $P(\mathcal{R}_r)$
s	the complex domain $s = \sigma + it$
t	$\operatorname{Im}(s)$ but also used as a dummy variable
t_i	$\operatorname{Im}(s)$ when $\zeta(1/2 + it) = 0$, the $\operatorname{Im}(\rho_i)$
t_c	value of t at a cycloid-like curve where the velocity will be zero
$[x]$	nearest integer to x, $\lfloor x \rfloor$ and $\lceil x \rceil$ floor and ceiling functions
$\mid$ and $\nmid$	$l\mid m$ means l divides m whilst $l \nmid m$ means l does not divide m

Pathway notation: a summary

Table B1 summarises the pathways of Euler's *zeta*, the modified Dirichlet's *eta* series and Dirichlet's *beta* series. A *distal* superstructure has paired *pseudo-spirals* with a *principal-axis* upon which a *proximal* vector from another pathway may be superimposed.

Table B1. Pathways, designated vectors and associated *distal* superstructures.

Pathway	Guide to structure	Vector	Set of vectors and Superstructure	Distal/Proximal	Principal-axis of superstructure relates to
Euler's *zeta*					
$P(\zeta_n(s))$		$\vec{m}$	$\mathcal{R}_r$ and $P(\mathcal{R}_r)$	κ	$\vec{\mathcal{R}}_r$
$P(h_{1,r}(s)) \equiv P(\bar{h}_{1,r}(s))$	$q_{\vec{r}} \in \chi_1 \equiv \mathbb{E}_2$	$\vec{\mathcal{R}}_r$	$\mathcal{M}_m$ and $P(\mathcal{M}_m)$	κ	$\vec{m}$
Modified Dirichlet's *eta*					
$P(\eta_{l,n}(s))$	$\varrho_{\vec{r}} \in \mathbb{E}_2$	$\vec{m}$	$\mathcal{R}_r$ and $P(\mathcal{R}_r)$	κl	$\vec{\mathcal{R}}_r$
$P(h_{l,r}(s))$	$q_{\vec{r}} \in \mathbb{E}_2$	$\vec{\mathcal{R}}_r$	$\mathcal{M}_n$ and $P(\mathcal{M}_n)$	κ	$\vec{\mathcal{L}}_n$
$P(\bar{h}_{l,r}(s))$	$q_{\vec{r}} \in \chi_l$	$\vec{\mathcal{R}}_r$	$\mathcal{M}_m$ and $P(\mathcal{M}_m)$	κ	$\vec{m}$
$P(\ell_{l,n}(s))$		$\vec{\mathcal{L}}_n$	L_n and $P(L_n)$	κ	$\vec{\mathcal{R}}_r$
Dirichlet's *beta*					
$P(\beta_n(s))$		$\vec{m}_\beta$	$\mathcal{R}_r$ and $P(\mathcal{R}_r)$	κ	$\vec{\mathcal{R}}_r$
$P(h_r(s))$		$\vec{\mathcal{R}}_r$	$\mathcal{M}_m$ and $P(\mathcal{M}_m)$	κ	$\vec{m}$

The vector $\vec{m}$ has no index because m is its own index.

The set L_n and the pathway $P(L_n)$ are mirror images of $\mathcal{M}_n$ and $P(\mathcal{M}_n)$ from $P(h_{l,r}(s))$. The latter are to be distinguished from the $\mathcal{M}_m$ and $P(\mathcal{M}_m)$ which are related to $P(\bar{h}_{l,r}(s))$. For *beta* only two complementary series are needed.

References

1. Mueller, I. *Philosophy of Mathematics and Deductive Structure in Euclid's Elements. Book Title*, 3rd ed.; Dover Publications, Inc.: Mineola, New York, 1981; pp. 58-83.
2. Edwards, H.M. Riemann's Zeta Function, The Dover edition.; Dover Publications, Inc.: Mineola, New York, 2001; pp 9-16.
3. The Experts Speak for Themselves. In The Riemann Hypothesis; Brownie, P., Choi, S., Rooney, B., Weirathueller, A. Eds.; Springer: Canada, 2008; Chapter 12.3, pp. 199–221.
4. The Experts Speak for Themselves. In The Riemann Hypothesis; Borwein, P., Choi, S., Rooney, B., Weirathueller, A. Eds.; Springer: Canada, 2008; Chapter 12.4, pp. 222–295.
5. Riemann, B. Ueber die Anzahl der Primzahlen unter einer gegebenen Grösse (On the number of primes less than a given quantity). Monatsberichte der Berliner Akademie 1859
6. Empirical Evidence. In *The Riemann Hypothesis*; Borwein, P., Choi, S., Rooney, B., Weirathueller, A. Eds.; Springer: Canada, 2008; Chapter 4, pp. 37–44.
7. Hardy, G. H. Sur les Zéros de la Fonction $\zeta(s)$ de Riemann. C. R. Acad. Sci. Paris 1914; 158: pp. 1012–1014.
8. Speiser, A. Geometrisches zur Riemannschen Zetafunktion. *Math. Ann.*1935 vol. 110, issue 1, pp.514-521.
9. Bohr, N. Über die Serienspektra der Element", Zeitschrift für Physik. 1920; 2 (5) pp.423–478
10. Montgomery, H.L., The pair-correlation of zeros of the zeta function, *Proc. Symp. Pure Math.* 1974, *Volume 24*, pp 181-193 (Amer. Math. Soc., Providence, R.I.,).
11. Hiary, G.A.; Odlyzko, A.M. The zeta function on the critical line: Numerical evidence for moments and random matrix theory models. *Math. Comp.* 2012, vol. 81, no. 279, pp. 1723-1752.
12. Keating, J.P.; Snaith, N.C. Random matrix theory and $|\zeta(1/2 + it)|$, Commun. Math. Phys. 214 (2000), 57-89.
13. http://www.dtc.umn.edu/~odlyzko/zeta_tables/index.html
14. Siegel, C. L. (1932), "Über Riemanns Nachlaß zur analytischen Zahlentheorie", Quellen Studien zur Geschichte der Math. Astron. und Phys. Abt. B: Studien 2: 45–80, JFM 58.1037.07, Zbl 0004.10501 Reprinted in Gesammelte Abhandlungen, Vol. 1. Berlin: Springer-Verlag, 1966.
15. Coxeter, H.S.M. Kaleidoscopes: Selected Writings of H.S.M; Sherk, F.A., McMullen, P., Thompson, A.C. and Weiss, A,I.; Wiley-Interscience Publication, 1995, ISBN 978-0-471-01003-6

ABOUT THE AUTHOR

Anthony Lander is a Consultant Paediatric Surgeon at Birmingham Women's and Children's Hospital, Birmingham UK.

Comments and corrections are welcomed

t.lander@nhs.net   a.d.lander@gmail.com.

History of submissions for peer review

1/04/2017   MDPI ***Mathematics.*** Rejected. One reviewer said *"I'm impressed with the author's amazing effort…"* but considered it *"…just a summary of experiments"*. A pre-print of that submission remains on-line at www.preprints.org. By 26/3/18 there had been over 713 downloads and over 1142 views, but it had received no comments.

18/08/2017 ***The Journal of Number Theory.*** Rejected. A reviewer said that *"The paper contains a few interesting observations and some nice illustrations."*

25/02/2018 ***Communications in Number Theory and Physics***. An immediate rejection without being refereed.

25/03/2018 **MAYFEB** ***Journal of Mathematics.*** A version similar to this monograph was accepted after review by two number theorists who requested some modest revisions. I anticipate that this may appear online shortly. Copyright obliges me to say that "The original publication is available at WWW.MAYFEB.COM."

This monograph and my submissions to the journals are no doubt void of mathematical style and sophistication, however, no rejection to date has, as yet, addressed the pith or substance of the arguments themselves.

Anthony Lander   26th March 2018

Funding:	This work was not funded.
Author Contributions:	The author was the sole contributor to this work and is responsible for all errors in mathematics, spelling, grammar, formatting and layout.
Conflicts of Interest:	The author declares no conflict of interest — apart from this work having been more fun than tidying the garage.

Notes

Notes

www.ingramcontent.com/pod-product-compliance
Lightning Source LLC
Chambersburg PA
CBHW040218220526
45473CB00001B/30

Sampler Book 4, Ontario in Colour Photos, Saving Our History One Photo at a Time

Photography by Barbara Raué

Series Name: Cruising Ontario
© 2018

Sampling from several towns

Each photo I take that precedes a demolition, or a natural disaster such as a tornado or a fire, is meeting this aim of mine of Saving Our History One Photo at a Time. There are more than 100 towns already photographed which you can visit without moving from your comfortable chair in your living room. Dream about what it was like in those by-gone days. Dream about what it was like to live in a mansion like one of these. Where would you like to travel to next?

Cover: 704 Gustavus Street, Port Elgin

Table of Contents

Niagara-on-the-Lake, Ontario – My Top 5 Picks

Harriston and Clifford, Ontario – My Top 7 Picks

Neustadt, Ontario – My Top 7 Picks

Port Elgin, Ontario – My Top 9 Picks

Wingham, Ontario – My Top 9 Picks

Lucknow, Ontario – My Top 5 Picks

Conestogo, Bloomingdale and West Montrose, Ontario – My Top 12 Picks

Delhi, Ontario – My Top 5 Picks

Waterford, Ontario – My Top 8 Picks

Waterloo, Ontario – My Top 12 Picks

Windsor, Ontario – My Top 13 Picks

Amherstburg, Ontario – My Top 10 Picks

© 2018 by Barbara Raue - All the photos in this book have been taken with my cameras. I own the rights to them.

Niagara-on-the-Lake, Ontario – My Top 5 Picks

Known at various times as Butlersburg, West Niagara, and Newark, its first permanent settlers, Butler's Rangers and other Loyalist refugees arrived in 1778 when they began crossing from Fort Niagara to settle the west bank of the Niagara River. A town was laid out in a grid pattern of four-acre blocks and grew quickly, gaining prominence as the first capital of Upper Canada from 1792 to 1796. The town was captured by American forces on May 27, 1813; upon their withdrawal on December 13, 1813, the American forces burned the town.

Following Niagara's destruction, the citizens rebuilt mainly in the British classical architectural tradition, creating a group of structures closely related in design, material and scale. Spared from redevelopment, the town's colonial buildings eventually became one of its greatest resources. Beginning in the 1950s, residents rehabilitated and restored old structures, demonstrating an exceptional commitment to the preservation of local heritage.

The Prince of Wales Hotel is a historic Victorian hotel located at King Street and Picton Street. Built in 1864, the three storey 110 room hotel went by several names (Long's Hotel, Arcade Hotel, The Niagara House) and was renamed with the current name after famous guests The Duke of York (and Prince of Wales) and The Duchess of York in 1901. Queen Elizabeth II stayed at the hotel during her visit to the area in 1973.

6 Picton Street – The Prince of Wales Hotel established 1864 - Second Empire style, mansard roof, dormers, window hoods, dichromatic brickwork, cornice brackets, second floor balcony – Book 1